How Invention Begins

How Invention Begins

Begins

Echoes of Old Voices in the
Rise of New Machines

JOHN H. LIENHARD

OXFORD
UNIVERSITY PRESS
2006

OXFORD
UNIVERSITY PRESS

Oxford University Press, Inc., publishes works that
further Oxford University's objective of excellence
in research, scholarship, and education.

Oxford   New York
Auckland   Cape Town   Dar es Salaam   Hong Kong   Karachi
Kuala Lumpur   Madrid   Melbourne   Mexico City   Nairobi
New Delhi   Shanghai   Taipei   Toronto

With offices in
Argentina   Austria   Brazil   Chile   Czech Republic   France   Greece
Guatemala   Hungary   Italy   Japan   Poland   Portugal   Singapore
South Korea   Switzerland   Thailand   Turkey   Ukraine   Vietnam

Published by Oxford University Press, Inc.
198 Madison Avenue, New York, NY 10016
www.oup.com

Oxford is a registered trademark of Oxford University Press

Library of Congress Cataloging-in-Publication Data
Lienhard, John H., 1930–
How invention begins : echoes of old voices
in the rise of new machines / John H. Lienhard.
p. cm. Includes bibliographical references and index.
ISBN-13: 978-0-19-530599-9
ISBN-10: 0-19-530599-X
1. Inventions—History.
I. Title.
T15.L485 2006
609—dc22
2005030825

9 8 7 6 5 4 3 2 1
Printed in the United States of America
on acid-free paper

Contents

Contents

Part IV
Views Through a Wider Lens

Preface

The narrator/ringmaster in the musical *The Fantasticks* begins the show by introducing its characters. Then, as he prepares to put them into action, he utters a wonderful line. "You wonder how these things begin," he says. "Well, this begins in a glen. It begins in a forest where woodchucks woo and vines entwine like lovers."

Beginnings of the important things in our lives are like that. They are quiet and invisible—not like the band-accompanied launching of a ship or the firing of a rocket into space. The ship might have begun with a student sitting on the rocks above the sea, hypnotized by the movement of a boat in the cove below or reading stories of the sea. Perhaps the Saturn rocket can be traced back to a child who watched Fourth of July fireworks, or read about Buck Rogers and Flash Gordon, and then asked, "Why not me?" Invention has fermented alone inside their heads, but it has also been driven by the voices of their communities.

We eventually realize that an elusive a priori essence hovers over all invention; we sense its presence, but it can be fiendishly hard to trace. Too much has always gone on before we get around to assigning priority to the creation of any new thing. Invention is a powerful part of the human psyche. But its texture and form are quite different from the cartoon images that we often use to represent it. We all want to foster the creative improvement of our world, and that alone is reason to spend this time sorting out the meaning of invention.

It is one thing to make the fairly obvious assertion that human invention is ever-present and that it is always accompanied by a communal synergy of ideas. To learn what invention is truly made of, we need to do more than just make that bland assertion. We also need to connect, viscerally, with its seeming contradiction—the coexistence of individual creativity and communal reinforcement. If we fail to ingest the contradiction, we can easily be tricked by a simplistic reading of the law of the excluded middle—the either/or requirement that disallows multiple explanations for one result.

The fabric of causality becomes terribly complex in the case of invention. That is why we do better if we begin with a seemingly illogical acceptance that invention is the emergence of a collective idea at the same time as it is an expression of one person's genius. Once we make that willing suspension of common sense, we are in a position to start looking for ways that the individual and the community form two facets of a single cause.

To undertake this process, let us spend a certain amount of time on the pathway of anecdote in preference to the straight highway of exposition. I ask you to join me in tracing folklore and history—to play with the mosaic of stories until a picture emerges from scattered tiles. Let us allow invention to reveal itself in very much the same way as it reveals itself to any inventor—by mutating from a jumble of ideas into a whole.

As we pursue this path, we shall (insofar as I am successful) come to better appreciate the vast sequences of invention that produce whole technologies. We see the word *invention* assigned offhandedly to such technologies as the airplane, the steam engine, the printed book, and more. I give away little of my story to say that, at the end, I need to propose a new word to describe these huge aggregates of invention. That way, we can reserve the fine word *invention* for the contributions that we all make repeatedly in daily lives.

Indeed, the aggregation of invention goes beyond even that. For example, we eventually realize that thousands of people applied their corporate inventive genius to something larger than airplanes, railroad engines, or automobiles. A collective desire, an upwelling of fascination and desire, a spirit of the times—a Zeitgeist—laid its hold upon them. The thing that they all sought to create was speed itself.

We likewise can hardly trace the astonishingly complex technology of printing books without coming at last to that which we desire from books—the knowledge, the *learning,* that books provide. Can we speak of speed or education as inventions? I suggest that it is no more of a stretch to do that than it is to call radio or the telephone an invention.

Undertaking these matters has been an ever-opening-up voyage of discovery for me, and so many people have been a huge help along the way. First in any such list must be my wife, Carol, who has read and commented upon draft after draft of each chapter. I am grateful to art and architectural historian Margaret Culbertson, who read the full manuscript and provided a great deal in the way of background advice and sources of illustrations. Medical historian Helen Valier (and her mother) also read the entire manuscript and provided a detailed, and very useful, commentary. My thanks as well to a large number of people who read and provided advice on selected chapters: Joyce Derlacki, Steven Mintz, James Pipkin, and Andrea Sutcliffe, as well as three anonymous external reviewers.

My thanks to the good people at Oxford University Press: editor Peter Prescott for his ongoing support and cogent advice; his assistant, Kaity Cheng, for so smoothly shepherding the book through the many hurdles of the book-making process; production editor Helen Mules for the grace with which she turns raw copy into this final form; copy editor Sue Warga for her deft touch; and all the other production people who function so effectively in comparative anonymity. I have also extracted more ongoing advice than anyone is entitled to from my generous colleagues: Sara Fishman, Barbara Kemp, Catherine Patterson, Stephen Perkins, and Lewis Wheeler. All these, and too many whom, by failing to keep proper track, I have failed to name have joined in along the way. Finally, my thanks to the fine ongoing support of the staff of the radio station KUHF-FM Houston.

Now I ask you to join me as well. As our *Fantasticks* narrator says, "You wonder how these things begin." Well, since we must begin somewhere, we shall do so high on a mountain crest, 5,300 years ago . . .

January 10, 2006   John H. Lienhard

Part I

Priority and Apriority

1

Ötzi and Silent Beginnings

Ötzi slumps against the rock, exhausted. The cold blunts his pain, but only a little. He's ten thousand feet up, on a crest of what you and I call the Ötztaler Alps. For two days, everything has gone terribly wrong. He and two other hunters were looking for deer and wild goats in the mountains when they found they'd trespassed on the turf of a mean band of hunters from another valley. They've been caught in a murderous, drawn-out skirmish ever since.

His forty-six winters make Ötzi the elder of his group. He's a seasoned hunter and a skilled archer, but during their first fight he loosed all his good arrows and was able to retrieve only two of them. One was broken; the other had found its mark. (The man he'd hit was dying when Ötzi pulled it out of his chest.) Now he's left with one good arrow and a quiver of unfinished ones. He'd meant to take care of them along the way, but each night he had been too tired for the work of smoothing the shafts, feathering them, and adding points. It was stupid to venture out here at his advanced age. He's paying a bitter price for overestimating himself.

Worse yet, they killed one of his friends in the fight. He and his other companion were finally able to disengage and run toward the safety of higher ground. Today, they were ambushed by two more of the angry strangers. Ötzi got off one shot—his only shot—but all it did was graze the man in front. The

two closed in, and there was a brief close-quarters knife fight. Ötzi and his friend did enough damage to make them back off, but not without getting badly cut up. Ötzi's hands and torso are slashed and bleeding.

Once more they succeeded in breaking free of the fight. When they took off on the run this time, a strange thing happened. By some crazy logic of chance, Ötzi noticed his one precious arrow on the ground as they ran by. Without thinking, he paused, stooped, and picked it up. That stupid mistake cost him dearly. It made a stationary target of him. As he rose, he caught an arrow near his shoulder.

When they were far enough away to pause for breath, his companion tried to pull it out. But his friend was also hurt, and his fingers were clumsy. The shaft broke loose, leaving the stone point embedded in Ötzi's back. And as they started off once more, his friend stumbled—went down on his knees and doubled over. Only then did Ötzi realize that the fellow had been stabbed in the stomach. Ötzi wrestled him up on his back and set off with the man's arms over his shoulders and his feet bumping the ground. After a few steps, Ötzi felt him go slack. He lowered him to the ground and looked for breath. There was none. His breath had left him.

So now Ötzi is alone and still on the run, that arrow point is festering in his back, and he's up on this high crest, surrounded by only rock and snow. He's a lot higher than he'd ever meant to be. The last of the killers seem to've given up the chase, but Ötzi's exhausted, he's lost blood, and he can hardly move his arm. He calls upon the goddess of the cold moon to get him out of this mess. It's growing colder by the moment, and the terrible pain is giving way to immobility. His arm has become useless. He sinks to the ground. Maybe he can rest for a few breaths and think this through—decide what to do next.

He still has most of his gear—his knife, his clothing, his pack frame. He still has his bow, but only that one good arrow. He has his fine axe with its copper head and its lovingly polished handle of hard yew wood—much good it'll do him now. He's far from home and a serious storm is starting to howl around him. No choice but to wait it out, and he knows it's death to stop moving.

He struggles to stay focused, to keep his mind from drifting back to his village in the valley far below, but he's beset by seductive images of green fields—of warm air, heavy with smells of cow manure, grain, grape, and mint. His imagination, unbidden, flees the pain and tries to summon up the roaring charcoal fire where he smelts copper for his small traffic in axe heads.

The screaming voice of common sense tells him to wake up. Breathe, think, stay alive! But that voice now comes from far down inside some remote cavern of his mind. He feels the pain ebb as his mind finally lets go. He drifts off, no longer hearing the full fury of an ice storm howling about him. Long before the next sunrise, Ötzi is frozen solid and covered over with snow.

There he remains for 5,300 years. Then, at 1:30 p.m. on Thursday, September 19, 1991, Erika and Helmut Simon, two experienced German mountaineers, on their way down from the Finailspitze, have been lured off their path by the warmth of this particularly beautiful day. This is the warmest summer on record, and here, just south of the Austrian border, on the Italian side of the mountains, they find Ötzi's head and back exposed at last by melting ice.

The Simons have no way of knowing how old he is. A piece of modern ski strapping, lately discarded on the ice above Ötzi, now lies beside him, suggesting that his death was recent. Erika Simon looks at the delicate articulation of his ribs and spine, at the exposed portion of his five-foot-three body, and decides this must be a woman. They both assume they've stumbled across a hiker who died in some accident, maybe a decade or so before. Helmut Simon takes the last photo in his camera, and they hurry down the mountain to an inn where they can report the discovery.

The police are called, and the process of extracting Ötzi from the ice and learning who he really was begins. An astonishing window upon life in the late Stone Age in southern Europe now rapidly opens. That view has improved with each passing year.[1]

One might rightly wonder how much liberty I have taken with Ötzi's story. I would be first to admit that other scenarios might also fit the facts. And as scientists continue sifting the forensic evidence, Ötzi's last days and hours keep becoming clearer. We now know that the point on his one good arrow carries blood from two other humans. Blood from two more people shows up on his knife, and there's more blood on the back of his cloak. The knife blade is flint, not copper like his axe head. It's small with a sharp scalloped blade. Ötzi carried a sophisticated little flaking device to sharpen it.

We've learned that Ötzi had recently eaten vegetables, ibex, deer, and some kind of flatbread. The grains reveal that his home must have been near present-day Bolzano, Italy. Since his hair carries traces of arsenic, a by-product of copper smelting, he most likely did his own smelting and metalwork.

Ötzi overturns so many convictions about chronology. He lived eight hundred years before the Great Pyramid was finished. Until fairly recently, we thought that the most advanced metalworking was that of the Egyptians, who had only begun hammering the occasional metal nugget into useful shapes. We thought that smelting had begun later and that these skills had then taken a long time to reach Europe.

Now we find Ötzi not merely shaping lumps of alluvial metal but actually smelting ore into very pure copper and casting it into useful forms. Ötzi had clearly stepped across the threshold that separates the ages of stone and bronze.

His light gear reflects a remarkable across-the-board knowledge of materials and of means for putting them to use. He used some eighteen kinds of wood and other vegetation to make his clothing and tools. Ötzi and his people were extraordinary technologists by any measure. Take, for example, his shoes.[2]

A Czech shoe technologist, Petr Hlavacek, studied Ötzi's shoes and feet. The shoes proved to be startlingly complex. The leather on the bottom was from a bear and had been cured in a mixture of the bear's brains with fat from its liver. Deer leather formed the tops. The whole array had been mounted on a mesh of braided linden bark and bound together with calf leather. Straw was used for insulation and moss for a lining.

Hlavacek and a colleague made three exact replicas of the shoes, and five more pairs fitted to living people. They used flint to cut the material and bone needles to sew it. Finally, they tested their Stone Age shoes in the snowy mountain terrain during the first spring melt—the time of year when Ötzi died. The shoes served remarkably well. When the hikers waded through snowmelt water, they felt an initial sting of cold, but the inside immediately warmed back up. Traction was excellent, and the shoes offered no opportunity for blisters. Ötzi was, in many ways, better shod 5,300 years ago than you and I are today.

And so, when we meet Ötzi, we feel as though we have broken the seal of a medieval crypt and found an internal combustion engine inside it. Of course it has been said that any good scientific answer generates two new scientific questions. This window into life in the late fourth millennium BC certainly does that. And we're of two minds in our reactions: a part of us wants to follow each emerging question and see where it takes us, but another part of us wants to reach a conclusion.

Whatever you and I know about history, that knowledge probably first took shape when we were schoolchildren. And it is very hard to deal with inconclusive questions in our early classrooms. Teachers suffer constant pressure to traffic in the answers to questions. Very often learning is reduced to the selection of the right answer from a list of four choices. Who is Ötzi? He is: (a) a 5,300-year-old mummified corpse; (b) a villager who looked at the clouds one spring day and deemed it a good time for a hunting trip into the great mountains rising above his valley; (c) a craftsman; (d) brave, cowardly, mystic, rash, a good friend, Italian—or too old at the age of forty-six to be running about in the mountains.

Who can look at Ötzi without aching to know him? The very fact that he lies here, right at hand, makes him frustratingly mysterious. What was his real name? What did he and his friends talk about? But we want credit for the right answer, so we write down (a) and go on to some other question: When was archery invented? When was copper first smelted? At some point we give up and formulate canonical answers to such questions just to keep from floating away on a great mushy cloud of ambiguity.

So, while others work to learn who Ötzi really was, most of us try to encase his story in some kind of digestible cocoon. Just as we tell one another that Watt invented the steam engine, Edison the lightbulb, Bell the telephone; just as we recite the myths of a Renaissance age of enlightened learning and art and a medieval era of darkness and prejudice; just as we tell children that Washington chopped down the cherry tree, we are equally tempted to fit Ötzi's complex existence into a viable container.

Perhaps that is not all bad. Maybe we need to begin with some framework, no matter how shaky it is. We begin by saying, "America was founded by people seeking religious freedom," and that might not be a bad place to start. Later on, we can sort out the witch hangings or the appropriate roles of church and state. For openers, it might be enough to tell students about the discovery of Ötzi and leave his role as an agent of historical deconstruction for another day.

Deconstruction is, after all, not something we want to undertake for its own sake. On the other hand, sifting out a better understanding of how any new technology came into being includes, by the very nature of the subject, a dimension of pure celebration. Ötzi may be an instrument of historical revision, but (his grisly death notwithstanding) what we've learned about his life leaves me wanting to dance and sing.

Much more is at stake here than revising dates. Rather, it is about seeing our heroes in perspective. For heroes play an important role in our culture. Leslie Brisman begins a study of the origins of Romanticism with a soliloquy on our hero myths. He quotes Rousseau, who said:

> It is no light undertaking to separate what is original from what is artificial in the present nature of man, and to know correctly a state which no longer exists, which perhaps never existed, which probably never will exist, and about which it is nevertheless necessary to have precise notions in order to judge our present state correctly.[3]

Brisman uses Rousseau's analysis to warn us that when the history of our origins has vanished over some now inaccessible horizon, those lingering myths of origins still serve us. Not only do our myths help us to understand ourselves when history has been lost, they also reveal things about ourselves that transcend any existing record of history.

If Rousseau is correct, and I believe he is, we need to pay attention to our myths of invention, even if we have a fairly complete knowledge of our objective history. Something happens when we transform a supposed seminal inventor into a figure of mythical proportions. The texture of our kinship with the inventor made hero is radically different from our kinship with the inventor as a fellow toiler. Ötzi and Prometheus, taken as exemplars, may ultimately be one and the same, but those two identities serve us in radically different ways.

We meet many heroes in this book. Some are now the stuff of myth, others not. But Ötzi has suddenly surfaced as the oldest technologist known to us, and his corporeal presence forever shields him from being transmuted into mythology. For me, his magic lies in the way his anonymity and his presence merge in a great historical contradiction. He was almost certainly a fine creative technologist. Yet he is without a name, just as you will inevitably be. The flesh and form of his battered body remind us of our own tired bodies. We look at him and we recognize what we see.

A better look at any of our origins through a lens that combines both passion and calculated detachment inevitably brings us back to ourselves. If we understand the texture of past creative accomplishment, we greatly improve our own chances of recognizing and building upon our own unsung efforts to give our children a better world.

Among the impediments to gaining that view is the way we presently describe our technological past, no doubt. But before we can alter our textbooks, we first need to learn just how invention works as a process. The far more basic impediment is the deeply ingrained, and often deceitful, concept of *priority*. Before we travel further with the descendants of Ötzi, we must look at priority itself. Just how has it directed our thinking about human creativity?

We have mentioned the way in which we seem impelled to reduce the cumulative thinking of all the Ötzis who anonymously formed our world down to a few named individuals. Perhaps it is a mere shorthand means for sustaining narrative in the absence of knowledge. Perhaps, too, it is our need to be guided by heroes. Why should it be so hard to identify with the idea of cumulative creative contributions? Yet difficult it is, and if this need for naming individuals did not serve some human function, it would not persist the way it does.

We obviously cannot talk about invention and historical antecedence without first untangling our seemingly atavistic need to credit one individual for the work of many. If we are to talk about how our technologies are brought into being, we need to think first about what I should like to call the invention of priority.

2

The Unrelenting Presence
of Priority

The question I am asked most often is, "Who first invented this or that?" I can never seem to give the questioner any proper satisfaction. In fact, I shy away from the question just because I know that, after the conversation has spun around, it will end at last in frustration for both parties.

Perhaps it is worthwhile to place one such question on the table here, so you and I can watch how the search for the answer plays out. Allow me to offer an ostensibly frivolous question, one that is not apt to awaken anyone's passions, but which is actually quite revealing. Let us ask, "Who invented the *doughnut*?"

One person who gets prominent credit is Maine sea captain Hanson Gregory.[1] Gregory's ship was named *Frypan* because he fed his sailors fried cakes. They were deep-fried according to his mother's recipe. A problem with those otherwise delicious cakes was that their centers were often undercooked. In 1847, Gregory punched out the center of a cake so that all the dough that was being cooked lay near the surface. The result was a far more uniformly cooked cake.

That story is often told with embroidery about eating doughnuts during storms at sea, about punching the hole with a belaying pin, and so on. But the problem with Maine's claim is that Massachusetts rises to the challenge with an earlier and far more fanciful tale. According to their account, Indians found a Cape Cod Pilgrim woman deep-frying cakes in an outdoor pot. They decided they could frighten her off by firing an arrow so it would noisily strike the pot. Then they could steal the cakes. But the bowman's aim was high. Just as the woman reached over to put another cake in the pot, the arrow drilled a hole through it. The woman screamed and fled. When she did, she dropped the cake into the oil, and the first doughnut was cooked.

Needless to say, such claims do not end with Maine and Massachusetts. According to Vermonters, their native son Shadrach Gowallapus Hooper invented the doughnut.

If one expects a patent to resolve the question, one gets John Blondell's 1872 patent for a wooden doughnut cutter—hopelessly late in the game. Another cutter was patented in 1891, this one made of tin. But cutter patents do not address the conceptual leap upon which the doughnut rides in the first place. That's the understanding that the cooked surface area per unit volume of dough is improved by a hole.

Naturally, when one looks closely, one finds European antecedents. And Asian cooking includes a deep-fried doughnut called a *vada*, whose dough is made from lentils. Another solution to the problem of cooking a cake uniformly is far cleverer than the doughnut, and very old. The New Orleans beignet and India's poori are only two examples of cakes *hollowed out from the inside* rather than drilled through. In both these cases, steam pressure develops when a small amount of water boils inside the pastry.

The poori and the beignet are both instances in which the motive force of steam is harnessed to do a task. One might even call them antecedents of the steam engine (Chapters 4 and 5) since steam does work in inflating the dough like a balloon. As we sift all the means used to execute this pretty obvious but still ingenious principle, the question of identifiable invention crumbles in our hands like dry leaves.

And this is not because the example was frivolous. The stories surrounding modern machines to which we routinely apply the word *invention* are no different. Invention constantly eludes us by ramification—by stirring and blending with additional ideas. The doughnut/beignet makes a fine exemplar of the norm.

The stories of inventing the steamboat, telegraph, airplane, or lightbulb may be told with a straighter face, but in every case

Top: An Indian vada.
Middle: A New Orleans beignet uncooked on the *left*, and one deep-fried on the *right*.
Bottom: The interior of a freshly cooked beignet, hollowed out by vaporized water.

their structure is similar. We have assigned inventors to each of those technologies, and it might often seem that we did so by tossing the names of people who worked in these technologies into a hat and pulling one out. Fulton, Morse, the Wright brothers, and Edison all made huge contributions that served to consolidate each of their technologies. But to call any of them a primary inventor of their technology is a little like naming the inventor of the first doughnut hole.

The culprit here is that word *naming.* So let us take a moment to weigh priority in relation to naming. Naming is, after all, a near-mystical expression of power. The book of Genesis offers a very important suggestion of the power of naming, right at the beginning. In Chapter 2 of Genesis we read:

> And out of the ground the Lord God formed every beast of the field, and every fowl of the air; and brought them unto Adam to see what he would call them: and whatsoever Adam called every living creature, that was the name thereof.

God grants Adam dominion over the creatures. Adam's first expression of that power is the act of naming them.

Overtones of naming reach far beyond mere identification. The rite of baptism carries the idea forward in the Judeo-Christian tradition— the notion that parents are responsible for stipulating the name by which a child will be known to God.

Shakespeare retained all the mystery of naming when he made a direct connection between naming and the creative process. In *A Midsummer Night's Dream,* he puts these familiar, but no less remarkable, words in the mouth of Theseus:

> The poet's eye, in a fine frenzy rolling,
> Doth glance from heaven to earth, from earth to heaven;
> And as imagination bodies forth
> The forms of things unknown, the poet's pen
> Turns them to shapes, and gives to airy nothing
> A local habitation and a name.

To create—to bring a new thing into being—is, in other words, to add a new name to our world.

The power of naming was given an important boost by a primal technology that was emerging at the same time that Ötzi lived. Rudimentary hieroglyphs were appearing in Egypt, far from the Tyrol, in the centuries that bracket Ötzi. Ötzi himself carried some fifty-nine tattoos on his body—all dots or straight lines arranged in abstract groups. These may have had something to do with healing, and in that role it is very likely that they contain information of some sort and are not pure decoration.

In fact, the dating of the first hieroglyphic writing depends, in part, upon what level of pictogram we are willing to accept as having made the leap from art into building blocks of narrative.

Means for recording words and concepts were forming, along with our ability to preserve both the identity and the context of real people. The written record would now provide human identity in perpetuity. The old oral traditions would continue turning remembered people into metaphor. I suppose they still do today. But the written word was a counterbalance. When we created writing, we turned legend into history.

So, who do you suppose the first person in this new written history was? Well, it turns out to be a highly creative inventor! Historian Will Durant offered Imhotep as the first real person who is well enough documented in Egyptian and Greek texts to take on flesh and blood.[2] Beginning with Imhotep, we leave the cardboard figures of legendary kings and patriarchs and encounter a human figure about whom we know some details.

Let us meet this Adam of recorded history. Imhotep was the advisor to the Egyptian king Zoser, who ruled around 2680 BC. Zoser was the dominant king of the Third Dynasty and the first ruler of what we call the Old Kingdom. The Old Kingdom itself is so named because it marks the point at which written records first begin reporting Egyptian history in any detail.

Limestone bas-relief of Imhotep (or Hotepa).

Imhotep was born near Memphis. We even have the names of his parents: Kanufer and Khreduonkh. He was known to be very wise, and he became a minister in Zoser's court—a vizier, chief ritualist, and counselor. He was also an architect. The heroic stone structures of ancient Egypt began under Zoser. Just a few centuries later they would culminate in the Great Pyramid of Khufu at Giza. Writing reveals just enough for us to understand that the force behind all that was not Zoser but Imhotep. Imhotep created a new architectural order.

He oversaw the building of the Stepped Pyramid at Saqqara—the first of the great Egyptian pyramids. It is the oldest of those heroic architectural treasures still standing. The Stepped Pyramid rises like a great wedding cake, and the ruins of what was once a delicate, four-acre, low-lying limestone temple surround it.

Imhotep's architecture is the part of his legacy that looms largest in our minds, because we can still see parts of it, but he was also a writer and a poet. However, his priestly role definitely subsumed medicine at the time. And here the mischief of priority assignment already touches Imhotep. For Egypt honored him more for medicine than for either writing or building.

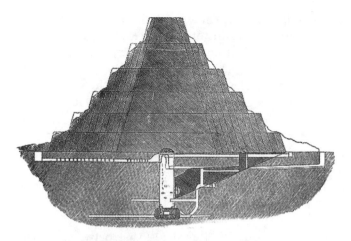

Cross section of the Stepped Pyramid (from *History of Architecture in All Countries*, 1883).

We have no idea what his medical contributions were, or even if he made any. We know only that the Egyptians eventually deified him for his healing. It is possible that by the sixth century BC Imhotep had displaced the god Toth as the god of healing, and his real father, Kanufer, was replaced with the Memphis god Ptah.

By then, the Greeks had their own god of healing, Asclepios. He too was derived from a real person whom Homer mentioned in the *Iliad* only as a fine physician. As Asclepios was expanded in the oral legends, he (like Imhotep) was deified. He was given Apollo for a father. Finally, Imhotep and Asclepios appear as a single god called Asclepios-Imhoutes. (Perhaps the Greeks were just hedging their bets.)

As a sidebar on all this, you may recognize the name of Imhotep not from any role in history but because his name was, for no particular reason, given to the title character in the 1932 horror movie *The Mummy*. The name Imhotep was resurrected (reanimated?) for the 1999 remake of *The Mummy* and its 2001 sequel, *The Mummy Returns*. And in 2002 we find it again in a bizarre but critically acclaimed comedy, *Bubba Ho-Tep*.

So although we might place the mantle of "first person in recorded history" upon Imhotep, his human person was eventually turned into something else entirely in our minds—god or movie monster, but no longer flesh and blood. History may be the worse for that, but history now existed where it had not existed before. Emerson would eventually write, "There is properly no History; only Biography." After Imhotep, we would have history and we would have biography.

Imhotep reached his bare threshold of history six hundred years after Ötzi was frozen into the Ötztaler Alps. The emergence of individuals as named historical figures would now mean that an occasional anonymous Ötzi would be turned into a historical figure. In that, the Stone Age did not so much end with the introduction of copper, brass, and iron—rather, it ended with written language. It ended with written names of real people.

Once we had names and identities, we used them to continue and elaborate the legends of greatness. You, I, and Ötzi all invent, yet none of us is ever really first to produce any technology. Our inventions always build upon antecedents. Why, then, do we buy into the concept of priority? I think the primary purpose for doing so is to celebrate the creative process. That is all well and good, so long as we do not let someone else's canonical greatness intimidate us—make us feel that our own inventive capability is too modest to put to use.

Today, we might expect to find that the linkage between names and inventions is firmly established by our system of patents. Indeed, when questions of priority arise today, all eyes turn to patents. Yet that linkage is far less solid than we might hope. While the concept of priority (shaky as it is) is embedded in patent law, patents are extremely unreliable identifiers of creative priority. Their real function is to sort out ownership in a world that feels it depends upon the ownership of ideas. Patent law has very little to do with the way invention actually works.

Indulge me for a moment while I offer an experience of my own with invention and patents.[3] In the 1970s, under contract with the Electric Power Research Institute (EPRI), I was trying to learn what would happen if a high-pressure hot-water line in a nuclear reactor fractured. The immediate problem was to simulate a pipe break by opening a high-pressure hot-water pipe very rapidly.

The two obvious options were bursting a metal diaphragm or blasting the end of the pipe off with explosives. Unfortunately, diaphragms don't tear fast enough, and tailor-made explosive charges are both messy and expensive.

I mentioned the problem to a colleague, who said, "Hmm, the pressure in that pipe would blow a cap off right now if you could just release it quickly." And I realized that we should be able to design a titanium plug combined with a guillotine device to accomplish a fast release. I sketched out the design for the graduate student who was working on the project, and together we built such an apparatus. With it we managed to drop the pressure in a pipe at the astonishing rate of well over a million atmospheres per second—far faster than anyone else had ever opened a pressurized pipe. In the end, the results of that work were translated into a part of the nuclear safety codes.

So EPRI filed for a patent on the device. When I saw the finished patent, I was astonished to find that lawyers had made fifteen different claims to creative invention from this one simple gadget.

Now EPRI owned a patent, which contained ideas (suggested by our work) that had never occurred to us. The credited inventors were the student and me. The colleague who had made the remark that triggered the patent in the first place had pretty much forgotten about the passing conversation.

The obvious question of just where what idea came from was now subsumed in the patent, and even as a primary participant, I cannot really claim to understand the precise evolution of the idea(s) now embodied in it. That's because I now understand how impossible it is to accurately parcel out individual credit for any invention. Our opening device was completely typical in being the ultimate result of ongoing communal creative input.

To see just how tenuous the patent is as a representation of priority, suppose I were to tell you that all the patent records over a half century had been suddenly and completely expunged. In that case, any new overlapping patent would have to be issued regardless of duplication. Where would priority go in that case?

Although that may sound like a crazy scenario, it is exactly what really did happen. Our U.S. patents are presently numbered sequentially, with a cumulative number that will reach 7,000,000 in 2006. These patents are issued by an office that opened in 1790, soon after the American Revolution. Yet patent number 1 was not issued until 1837, forty-seven years later.

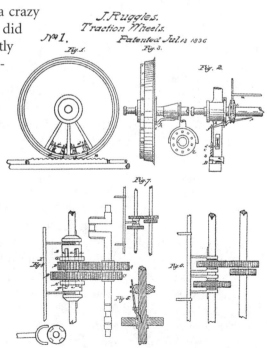

To see why, let us go back to the War of 1812, when the British burned Washington. The building that then housed the Patent Office and General Post Office was the only government building to survive. Congress met in that old building while the Capitol was built. The existing patents had survived the fires, and during the next twenty-four years, inventors continued adding to their number.

By 1836, however, the building had become hopelessly outdated and overcrowded. This one original building was now a firetrap and home to ten thousand patents, along with seven thousand patent models. As it threatened to burst its seams, the government finally authorized construction of a new office.

Just then, as if on cue and before construction began, a fire broke out in the old building. The new building had been authorized too late. The old Patent Office had survived the War of 1812, but now it succumbed to a mere accident. The fire destroyed the office and everything in it.

The Patent Office now had to be rebuilt from scratch. Only as the office began anew, did it initiate its present sequential system of numbering patents. Patent number 1 was issued to John Ruggles in 1837. It described an improved railroad wheel, meant to gain better traction.

The office managed to reconstruct two thousand of the ten thousand lost patents during the next decade. They assigned a parallel sequence to the resurrected patents, numbering them as X-1, X-2, and so forth. They attempted to place those numbers in the correct chronological sequence as well, and that has led to an odd complication. When a rediscovered patent is deemed to lie between two adjacent numbers, it is given a fractional value so the sequence will be correct. We've come to call those rebuilt documents the "X-patents."

The yield of lost patents slowed almost to a halt after about a decade. So the Patent Office abandoned active reclamation work, writing off the remaining eight thousand patents as an irretrievable loss of our national legacy.

Others, however, have continued the search. In 2004, *New York Times* writer Sabra Chartrand described the surviving process of patent archaeology.[4] Some eight hundred additional X-patents have been restored since the Patent Office gave them up for lost, and the work goes on. Bit by bit, the record continues to be reconstructed. And the task is being done by the most effective workforce in the world—dedicated amateurs. Some of those "amateurs" are seasoned patent attorneys, but they are doing the work on their own, pro bono.

The trigger for the *Times* article was one discovery in particular, and it is very revealing. Copies of fourteen old patents turned up in the library of Dartmouth College. Ten of them had been donated by Samuel Morey, an important early American inventor.[5] One of Morey's X-patents was already well known, at least in his native New Hampshire. Filed in 1826, it described a gas-driven internal combustion engine. That was long before the Otto cycle and probably it was the first gasoline engine—a stunning example of one of the inventions that undercut our canonical priority assignments.

Much earlier, Morey had patented a steamboat and actually built a working model. Morey tried to sell his X-patent for that one to Robert Livingston, the financier who ultimately backed Robert Fulton. Morey took Livingston for a ride on his boat, but the two could not come to terms on how to arrange the finances or how to divide the potential profits.[6]

When Fulton finally filed his own patent, he emphasized the paddle wheels on his boat. That was a part of Morey's steamboat that (unlike all those claims on my pipe-opener patent) had *not* been included in Morey's patent. The rest, as we say, is history.

Morey's presence among inventors of the X-patents suggests what their true meaning might be. They were written when American invention was still very young. Today, we have seven hundred times as many patents as we did in 1836, but those missing patents, written upon an almost clean slate, have a certain primacy over later ones. They undoubtedly include more potential nuggets of surprise—more treasures (such as pre-Fulton steamboats and pre-Otto internal combustion engines) that have been duplicated during the long years since.

They also dramatize the fact that almost nothing has any one absolute inventor. We recite the name of this or that lonely genius so often that we start believing such a person exists. Yet he is, by and large, a created hero who manifests the many builders of our present world all rolled into one person. He is the hero whom we invent to explain who we now are, in the absence of full knowledge of our past.

Therefore, let us not stop with doughnuts, pipe openers, and the supposed potential of undiscovered X-patents. We need to see how the process of unpriority—of *apriority*—works in our major technologies. We can pick any one out of a thousand, but this time let us begin with a particularly deceptive one.

Very few creations of major technologies trace as convincingly to a particular inventor as the airplane traces to the Wright brothers. But we would be hard-pressed to find another with as many competing claimants. For that reason, the airplane makes a fine place from which we might take a larger look at the strange business of assigning priority.

3

I Built My Airplane Before the Wright Brothers Did

The question of priority grew particularly pointed as the Wright brothers' centennial put it under a spotlight before the centennial celebration of their December 17, 1903, flight. During those months in 2003, we heard mounting claims from every quarter.

New Zealand made an energetic counterclaim. Great Britain has several plausible claimants. We heard from France and Brazil. Here, in the United States, we heard from Connecticut, California, Texas—indeed, it seemed almost craven not to enter a competing first flight from one's own geographical place.

I remain content to say that the Wright brothers were first to fly, as long as we saddle their claim with enough adjectives. They made the first powered, heavier-than-air, controlled, repeatable, unlaunched flight. Even then, various contenders have claimed many parts of that list, and others have faulted the Wrights on additional criteria. For example, Orville and Wilbur used launching rails at first. Do they really get credit for rising off the ground?

This is one case in which naming an inventor probably makes sense, simply because the Wright brothers so convincingly put so many of the pieces together in a way that no one else had come close to doing. They make an excellent candidate for what I would call the airplane's "canonical inventor." Yet they are, and will always remain, particularly vulnerable to counterclaims because so many people have, since time immemorial, struggled to make machines that would fly. And, down through millennia, many who have tried to fly have been partially successful.

The primal, even atavistic essence of our craving to fly came home to me in a strangely skewed way some years ago in Washington, D.C. One morning a historian whom I had been working with handed me an article from a magazine that he had found in a supermarket checkout line (you know the medium). It said, "Soviet researchers [find] an Egyptian mummy at the North Pole." The article claimed that ancient Egyptian explorers had flown to the North Pole in a human-powered airplane. This had been done to provide one of their honored dead with a special burial.

The man watched with a flickering smile while I read the article. (He could not have known that I had always dreamt of working as a headline writer for one of those magazines.) I would have forgotten the incident if it had not been for a strange revelation on my own airplane trip home. I had chosen for my "airplane read" a copy of Anne Rice's novel *The Mummy*.[1]

Up in the cold reaches of 36,000 feet I met her protagonist, a reanimated mummy. She gave her mummy the identity not of Imhotep but of Ramses II. So I read about Ramses walking the streets of early-twentieth-century Cairo. Returned to life after more than three millennia, Ramses is bewitched by modern technology. He cries, "There is so much to be discovered . . . we must go to the North Pole in an aeroplane."

Two mummies flying to the North Pole within less than two days got my attention. Then I began recognizing other echoes of mummies in the ice and in the air. There was Clive Cussler's novel *Treasure*.[2] It's about a ship from ancient Egypt found frozen into the ice of northern Greenland with its crew perfectly preserved. Earlier, Mary Shelley's Gothic tale of horror captured this mad imagery perfectly. It began and ended with a crazed Victor Frankenstein chasing his monster, his living mummy, across the Arctic ice.

Naturally our attention turns back to Ötzi. For here is yet another mummy frozen in the ice, a man who has at last completed an equally improbable flight across five millennia. One might almost wonder if, like Frankenstein's creature, Ötzi had not sought out the highest and coldest fringe of his own universe as a proper place to die.

The image of ice and death is ancient. Zoroastrians see the north wind as evil. In Hindu writings, Lord Siva had to behead his own son when he found him asleep, his head oriented to the north. The legends of the hot arid countries associate cold, and the north, with evil and finally with the end of life. Dante located the portal of Hell at the North Pole, and he described its first circle as a place where sinners are frozen up to their noses in ice. Dante also placed his Hell in the universe as it had been described by the Egyptian astronomer Ptolemy 1,300 years before.

Egypt and the North Pole: warmth and frigidity, life and death, being earthbound and being aloft—the themes intertwine. That crazy supermarket magazine had beautifully embodied all the old myths of contrast. Then it went on to couple them with our ancient and primal need to fly.

The myths represent something beyond mere flights of imagination. Sooner or later people act upon their most deeply seated cravings. The myths have to have arisen because, here and there in the recesses of the past, people tried to fulfill their desires. That was made especially clear in 1988 when a team from the Massachusetts Institute of Technology, fascinated by the myth of Daedalus, used twentieth-century technology in an attempt to replicate Daedalus's flight. They flew a human-powered airplane, which they named *Daedalus,* seventy-four miles northward from Crete to the island of Santorini.

Daedalus, now in the Museum of Science, Boston.

That was not as far as Daedalus's legendary flight from Crete to Sicily, but it was a large enough fraction of it to make us wonder if fragments of history might be embedded within the legend. Perhaps we should include the story of Daedalus among what I would call the transmythological stories—those tales that have been so elaborated by storytellers over the centuries that we begin mistaking truth for myth. Consider a case in point.

The summer of 2003 found American troops fighting at the Ibn Firnas airport, just north of Baghdad. Few Westerners gave particular thought to that name, or why it was attached to an airport. But Ibn Firnas is one of the earliest named and documented humans who really flew.[3] And still, so much storytelling surrounds him that he (like Imhotep) becomes separated from the actual record of his life.

Firnas's story began in the ninth century AD. All but a northern strip of present-day Spain and Portugal then formed the Andalusian Caliphate of Cordova. This was the high tide of Islamic art and science. Cordova and Baghdad were twin cultural centers of the world in what are now Spain and Iraq. In 822, Abd ar-Rahman II became the new emir of Cordova, and he set about to create a renaissance in the Moorish court there. His ingathering of talent began with an Iraqi musician called

Ziryab, which meant "blackbird"—a nickname that honored not any impulse to fly but rather his fine singing and dramatic appearance.[4]

A jealous music teacher had driven Ziryab out of Baghdad. So Abd ar-Rahman hired him at a handsome salary. In Cordova, Ziryab developed new musical forms. He introduced the lute to Spain and expanded its range by adding a fifth string. But Ziryab also became a patron of the sciences. He fostered the development of astronomy, medicine, and many technologies. Into this exciting world, so bubbling with ideas, came Abbas Ibn Firnas, a young Berber astronomer and poet from North Africa.

In 852, a new emir, Muhammad I, took power and a bizarre experiment took place. A daredevil named Armen Firman publicly leapt off a tower in Cordova, breaking his fall back to earth with a huge winglike cloak. He survived with minor injuries, and the young Ibn Firnas was there to see it.

Like Ziryab, Ibn Firnas worked at a huge variety of enterprises. He set up astronomical tables, he wrote poetry, he built a planetarium, and he designed a water clock. He also developed a process for cutting rock crystal. Up to then, only the Egyptians knew how to facet crystal. Firnas's process made it possible to work more of the caliphate's quartz at home, instead of sending it off to Egypt to be finished.

Yet Firman's flight must have lain upon Firnas's mind, for in 875 Ibn Firnas built a human-carrying glider. His machine was far more than an umbrella-like cloak, and in it Firnas also launched himself from a tower. The flight was fairly successful, although the landing was bad. He injured his back and left critics saying that he had not taken proper account of the way birds pull up into a stall and land on their tails. He had provided neither a tail nor means for such a maneuver.

His death twelve years later may have been hastened by the injury. And, just as we tell our schoolchildren about the Wright brothers, many in the Islamic countries tell theirs about Ibn Firnas, more than a thousand years before the Wrights. The Libyans have a postage stamp honoring him, and the Iraqis have their airport.

It is likely that word of Ibn Firnas's flight reached England, since we find that a surprisingly similar attempt to fly was made just after the year 1000 by a monk named Eilmer at Wiltshire Abbey.[5] The outcome was also strangely similar to Firnas's. Eilmer lost control, broke his leg, and subsequently said he should have had a proper tail on his machine. And he too was lame for the rest of his life.

We will never know how many people have tried to glide off towers or return to earth with the help of primitive parachutes. But countless attempts to fly certainly occurred long before history took note of them. Many Chinese, for example, flew in kites, some willingly, some not. Down through the millennia people have swung on vines, dived into water from high cliffs, tobogganed, and ski-jumped. Whether we rode a vine or a roller coaster, we have, in every age, found means for creating illusions of flight.

Throughout history, we've watched the birds wheeling and turning against the backlight of the sky, studied the delicacy of their movements, and dreamt of feeling that same kinesthetic joy. We have shared an elemental craving to join them in the sky. Sometimes we have even found means that took us there, if only for a moment. Indeed, when the French learned to leave the earth in balloons, in 1783, it was almost anticlimactic.

By the turn of the nineteenth century the new analytical sciences began embracing the dream of flight. Sir George Cayley was a prime exemplar of this new breed of inventors, grounded in the tools of mathematical analysis.[6] Born in 1773 in Yorkshire, he was studious from the start, and determined to solve the old riddle of human flight. He was ten years old when the French invented hot-air balloons.

But Cayley knew that serious flying machines would eventually have to be heavier than air. By his early twenties, he'd built a laboratory at his ancestral home of Brompton Hall and was undertaking sophisticated aerodynamic studies. He also hung out in a local watchmaker's shop—studying machinery at the same time he read Newton.

Cayley made several important discoveries. He realized that the secret of flight was to be learned not from birds' flapping wings but by watching birds glide with their wings fixed. He identified the three forces acting on the weight of any flying object—lift, drag, and thrust. To the best of our knowledge, it was he who conceived the idea of creating lift by giving a wing an airfoil-shaped cross section.

In one remarkable burst of insight, Cayley saw that trout have the ideal, minimum-resistance, body shape for an airplane. Why a trout and not a bird? A century later the new theory of dynamic similitude would show how to compare bodies of different size, moving at different speeds in different media, when forces of inertia and viscous resistance act upon them.

Cayley realized, long before that theory was in place, that the movement of a small fish in water is similar to that of a large machine in the air, while the flight of a small bird in the air is not. By 1799 he had formulated the basic shape of a modern airplane with a fuselage, wing, and tail. By 1804 he was flying model gliders. In 1809 and 1810 he published a series of articles on his theories and experiments.

He also showed how to resize his models in such a way as to make controllable human-bearing gliders. Cayley scrupulously backed everything up with calculation. Newton had proposed a simple theory of lift, which Cayley correctly took to task. (Newton's theory makes sense only at speeds far greater than sound. Ironically, it had no practical value until after World War II.)

Cayley brilliantly analyzed two of the three forces that must act upon the airplane's mass to accomplish aerial navigation: lift and drag. The third force was thrust, and in Cayley's time the only machines with any possibility of thrusting an aeroplane forward were large steam engines—far too heavy to carry in an aeroplane. So he put aside his studies and took up Whig politics. He became a member of Parliament.

But others read his principles (while steam engines became more efficient). By midcentury the aging Cayley could no longer dodge his legacy. In 1853 he built a full-size glider and ordered his coachman, John Appleby, to test it. An eager crew towed it into the sky, and Appleby glided safely down a hill.

Appleby, however, was no kin to Ibn Firnas or Daedalus. His fear overpowered any sense of history or elemental cravings that he harbored. He immediately tendered his resignation. Cayley died soon after that, without ever creating what might well have received credit for being the first successful powered flight. The studious and scientific Wright brothers read Cayley's work and were, by then, able to build an aluminum internal combustion engine. Few of us remember Cayley, even though his ghost rode along with the Wrights at Kitty Hawk.

Cayley was followed by other inventors, equally immersed in the analytical underpinnings of flight: Penaud, Giffard, Stringfellow, Lilienthal, Maxim, Chanute, Langley, Santos Dumont—I would not want to be held responsible for constructing a complete list, much less for telling all their stories. A vast number of people turned the focused human intellect upon the innate, almost animalist urge to fly before the Wrights did, and many of them actually got off the ground in the process.

Let us illustrate this process with two very different American loci of the late-nineteenth-century impulse toward flight. We find one on a side trip to Texas at the turn of the twentieth century—an obsession headed down a forgotten road that led nowhere. The other is quite different in texture. Centered in the then backwater of early California, it gains in momentum, forms its own infrastructure, and reaches out to become a part of the mainstream of our gathering effort to fly. Neither by itself is representative. It is only when we take them together that they help us to understand the birth pains of flight.

First, our side trip into the obsessiveness typical of so many pre-Wright would-be airmen. Writer Michael Hall tells how fifty-two-year-old Reverend Burrell Cannon decided, in 1900, that he was being called by God to replicate Ezekiel's wheel.[7] Cannon, as much a businessman as a preacher, was running a lumber mill in the east Texas town of Pine. But he'd spent fifteen years studying the book of Ezekiel, and he set out to build an Ezekiel wheel flying machine based upon that biblical account.

Let us look at the actual text, using the same King James Version that Cannon read. Let us discover what a daunting task it must have been to make scripture into a design manual. (You might well be tempted to skip over this remarkably opaque passage, but I suggest that you try to see it through the eye of Cannon's mind. It is a text you are unlikely to have encountered in any usual Sunday/Sabbath Bible reading.)

> Now as I beheld the living creatures, behold one wheel upon the earth by the living creatures, with his four faces. The appearance of the wheels and their work was like unto the colour of a beryl: and they four had one likeness: and their appearance and their work was as it were a wheel in the middle of a wheel. When they went, they went upon their four sides: and they turned not when they went. As for their rings, they were so high that they were dreadful; and their rings were full of eyes round about them four. And when the living creatures went, the wheels went by them: and when the living creatures were lifted up from the earth, the wheels were lifted up. Whithersoever the spirit was to go, they went, thither was their spirit to go; and the wheels were lifted up over against them: for the spirit of the living creature was in the wheels. When those went, these went; and when those stood, these stood; and when those were lifted up from the earth, the wheels were lifted up over against them: for the spirit of the living creature was in the wheels. And the likeness of the firmament upon the heads of the living creature was as the colour of the terrible crystal, stretched forth over their heads above. And under the firmament were their wings straight, the one toward the other: every one had two, which covered on this side, and every one had two, which covered on that side, their bodies. (Ezekiel 1:15–23)

That's cold comfort to any engineering designer, but Cannon parsed it as best he could.

Cannon was businessman enough to know he would need money and backers. So he sold his mill and moved to the nearby cotton center in Pittsburg, Texas. There he preached both the Gospel *and* his aeroplane. The times, says Hall, were optimistic, and Cannon found backers. Cannon sold $25,000 worth of stock, starting at $25 a share, and began building a one-man 26-foot prototype. His ultimate goal, however, was far more ambitious. He meant to create a 125-foot machine capable of carrying twenty tons. Using God's own design, how could he fail? Cannon did indeed finish building his prototype aeroplane.

It was an almost-circular flying wing, with a secondary lower wing below. The supporting structure was a light tubular metal frame. To match the biblical description of "a wheel within a wheel," Cannon created two pairs of wheels, nested below the wings. Since the Bible said that "the spirit of the living creature was in the wheels," Cannon placed his 80-horsepower engine (his own animating spirit) between the wheels.

Each wheel in the outer pair was eight feet in diameter. They taxied the airplane up to takeoff speed. The inner pair of wheels was smaller and faster-moving. They carried a pair of paddle wheels, meant to drive the machine once it was airborne.

That sets off alarm bells. Paddle wheels work on a steamboat because they push against the water at the bottom, then coast around through the air on top. Here the wheels are entirely in the air. Cannon managed to get around that problem (and remain true to the book of Ezekiel) by devising means for retracting the paddles in the upper return stroke. He also contrived means for controlling the machine in flight by varying the speeds of the two paddles.

Finally, in 1902, one of Cannon's workers flew the machine. It gained speed, took off, seemed to drift in the air, began vibrating violently, and then crashed into a fence. The flight, if you can call it that, had covered a distance of 167 feet.

Cannon lost his backers and moved back to Pine, Texas. But he was not done yet. He kept his original airplane and nine years later was at it again—this time with more backers and another airplane. That new plane crashed into a telephone pole. Hall ends his surreal story of Cannon and Ezekiel with a marvelous paraphrase: "Cannon's reach exceeded his grasp—by an inch or by a mile. It doesn't matter. What's a heaven for?"

Now, over and against the strange tale of Burrell Cannon, let us weigh our California story. Though it, too, has its wacky dimensions, it differs from Cannon's story in many ways. It has a lot more players, and they do not stay isolated from the flow and flux of the large external world. In fact, this tale does not even begin in California. Rather, it starts with a promoter and inventor named Frederick Marriott in Great Britain.[8]

Marriot had been a publicist for William Samuel Henson's and John Stringfellow's failed attempt to build a steam-powered airplane during the 1840s.[9] They, in turn, had been mentored in their almost successful efforts by Cayley, and Marriott had coined the word *aeroplane* in promoting their machine.

Sometime around 1848, Marriott had moved on to America and to the newly discovered California gold fields, hoping that his ambitions would fare better in a new land. There he was soon involved in a large number of enterprises. He became a San Francisco banker and a newspaper owner. He was a friend of Mark Twain, and one of Twain's early publishers.

By 1866 he was financially solid enough to form the Aerial Steam Navigation Company. Its immodest aim was to create steam-powered airship service between New York and San Francisco. (By then, only one such airship had ever flown. Jules Henri Giffard demonstrated his little one-man dirigible in Paris, in 1852.)

Marriott didn't try to do it all in one bite. First he made a 37-foot unmanned dirigible. It looked like a big football fitted with control surfaces. Two propellers were driven by a beautiful little alcohol-powered steam engine, only a foot long. He named his airship *Hermes Jr. Avitor*. His plan was to follow it with the *Hermes Avitor*—a 150-foot manned dirigible.

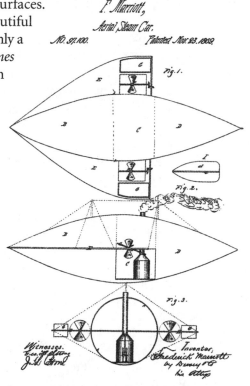

During the summer of 1869, he repeatedly flew his *Avitor* in San Francisco. A ground crew would run after the airship setting its controls with a rope. It was clearly successful at doing as much as it had been designed to do, and Marriott rhapsodized about its potential: "No savages in war paint," he said,

shall interrupt its passage ... across our continent. No malaria, or hostile tribes nor desert sands shall prevent the exploration of Africa. ... Man rises superior to his accidents when for his inventive genius he ceases to crawl upon the earth and masters the realms of the upper air.

Frederick Marriott's dirigible patent drawing.

Marriott was set to build the large *Avitor* when disaster struck his company in the form of the 1870 stock market crash. By 1884, Marriott had recovered from his financial setback. His dream of flight was still alive, but now he turned his attention from dirigibles to building an aeroplane. He might have produced something very interesting indeed had he not died later that same year at the age of seventy-nine.

And that might have been that. However, Marriott had a young nephew, John Montgomery. The eleven-year-old Montgomery had watched the *Avitor* in flight, and his uncle Frederick built him a small, wheeled model of the *Avitor* to play with. John Montgomery was clearly bitten by the flying bug, and it is to his part of the story we turn next.

Montgomery was born in 1858 in Yuba City, California, a scant ten years after the California gold rush and not far from the site of Sutter's Mill.[10] As he grew up, he watched birds and thought about the means they used to fly. His parents were influential citizens, and he received the best education available in that remote world. He attended several of the new private schools and colleges that were springing up in the wilderness and helping to transform it. He finally finished both a B.S. and an M.S. in physics at St. Ignatius College in San Francisco.

After college, Montgomery went back to the family ranch (now located near San Diego) to learn how to manage it, as well as to set up his own physics lab and shop on the property. Uncle Frederick's influence clearly lingered, because now Montgomery began developing his own flying machines. His first attempts were model gliders whose wings were flat—without the cross-sectional airfoil shape that successful airplanes would use. Without the appropriate curvature on the top and bottom of the wing (that is, without camber) they were failures.

It is at this point that Montgomery's story becomes tricky to sort out. We essentially have two versions, one by Wright brothers historian Tom Crouch, the other by Montgomery supporters.[11] When the two are held up to the light, the differences are fairly minor, yet they generate a good deal of factional heat.

Montgomery began building wings curved in cross section. Montgomery later insisted that the idea of a cambered airfoil, inspired by watching birds in flight, was completely original with him. Perhaps he had read Cayley or heard about him from his uncle Frederick Marriott, perhaps not.

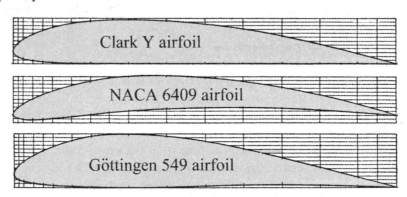

Some standard forms of cambered airfoil sections as they finally evolved before World War II.

In any case, Montgomery made additional attempts at glider building. Then, in 1883, the same year his uncle died, Montgomery built a fixed-wing glider equipped with cambered wings. In it, he made a successful flight, and here the waters grow muddy. He laid off glider building and remained secretive about all he had done.

Finally, in 1893, Octave Chanute organized a conference on flight in Chicago (coincident with the great Chicago Columbian Exposition). Montgomery attended, and there he told Chanute that he alone was the one person who had really flown.

Chanute was a leading civil/mechanical engineer—a former chief engineer for the Erie Railway—with wide-ranging technical interests. As early as 1885, Chanute had organized a session on flight at a meeting of the new American Society of Mechanical Engineers. By 1893, he was a major clearinghouse for information about flight. The Wright brothers leaned heavily upon him in their early work; he even visited them to consult on their glider experiments in North Carolina.[12]

For many years, Montgomery's story has been largely told according to Chanute's retelling of it in his conference report. Montgomery did not write the report, but he supposedly read and approved Chanute's proofs. In Montgomery's detailed account, he flew 600 feet. In Chanute's record, he flew only 100 feet.

During the next few years, first Otto Lilienthal in Germany, then Octave Chanute in Ohio, and finally the Wrights in North Carolina all went on to make repeated and completely successful glider flights. (Chanute was sixty-four at the time and did not try to pilot his own gliders.)

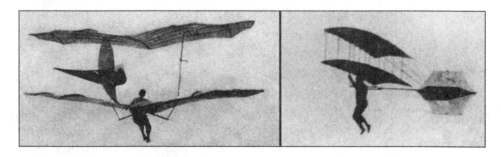

Left: Otto Lilienthal aloft in the same glider in which he crashed and died in 1896. *Right*: One of Octave Chanute's gliders, probably in the same year.

After the Chicago meeting, Montgomery returned to glider making, doing a great deal of trial-and-error work. The Wrights had meanwhile learned to combine both a cambered airfoil (which they'd developed in their own wind tunnel) and a wing-warping system for controlling their flying machines. When they finally flew under power in 1903, Montgomery seriously began to chafe.

Nevertheless, he made one more significant achievement in 1905. He built his *Santa Clara* glider with two tandem cambered wings. Instead of flying it down a slope, he arranged for it to be dropped from a balloon. The *Santa Clara* was carried to an altitude of 4,000 feet and released. A very brave pilot named Daniel Maloney spent the next fifteen or twenty minutes riding it on its lazy journey back to earth.

Trouble appears to have intensified after that. The combination of hyperbole on the part of Montgomery supporters and minimization of his accomplishment on the part of Wright supporters has, over the years, made objective exploration of Montgomery's accomplishments very difficult.

Montgomery wrote a book on flight in 1909. He also found a champion named Victor Loughead, who wrote two books about flight and kept adding to Montgomery's legend.[13] Loughead portrayed Montgomery as one of the great theoretical thinkers of all time and the inventor of airplane controls. No doubt Montgomery's use of control surfaces showed significant foresight. Yet by the time Loughead was done, Montgomery had practically invented the airplane. And people kept trying to nourish that claim. Hollywood actually made a movie about Montgomery in 1946. The film *Gallant Journey* starred Glenn Ford as John and Janet Blair as his wife, Regina.

Victor Loughead, however, brings another dimension into the California impulse toward flying. His mother, Flora Haines, was born in Milwaukee in 1855.[14] She graduated from Lincoln University in Illinois and began writing for various western magazines and San Francisco papers. She also married a Californian and moved to San Francisco, where she wrote her first book, *Libraries of California* (now a collector's item). But the marriage ended. In 1886 she wrote *Handbook of Natural Science* and then she married John Loughead. After that, she became a prominent fiction writer, producing books that carried such titles as *The Man Who Was Guilty* and *The Black Curtain*.

Her marriage to Loughead also ended, and she married a third time when she was fifty-three. Her son by the first marriage, Victor (now using Flora's second husband's surname), became a noted American automotive engineer. Victor was also strongly interested in flight. Soon after the Wright brothers flew, he wrote his books on airplanes. And his younger half brothers, Allan and Malcolm Loughead, followed his interest. Allan became an automobile mechanic and then learned to fly. He did some early exhibition stunt flying—horribly dangerous work.

But looking at those hopelessly unsafe airplanes, Allan realized he could do better. He went back to San Francisco and talked Malcolm into going into the airplane-building business with him. Working part time, Allan and Malcolm Loughead produced a little seaplane in 1912—a biplane with one pontoon. That modest effort turned into a major airplane factory, making bigger and better seaplanes. They named the company after themselves and, in 1926, legally changed their own and their company's name from the old Scottish spelling to Lockheed.

Behind these three remarkable brothers was their mother, Flora Haines Loughead, who appears to have been a driving engine of discontent. She was in and out of marriages and always writing. An 1898 article in a Catholic magazine tells us something about her. She writes about San Francisco's first cathedral, St. Mary's Church, which the Jesuits had abandoned in a neighborhood near Market Street. Paulist fathers made it a base for work among the poor. They taught English to Chinese laborers while the neighborhood slowly died of crime and prostitution. Flora Loughead points her accusing finger at city hall with barely bridled fury:

> Today this fight of good-citizenship and decency against immorality and corruption is one that occupies the minds of thinking people in San Francisco over and above the issues of the fall elections. The police have refused to act . . . without the passage by the Board of Supervisors of a special act of condemnation closing certain illicit thoroughfares; and these latter refuse to perform their plain duty.[15]

The year 1900 found both Flora and a younger woman, Harriet Quimby, writing feature articles for the *San Francisco Chronicle*. Quimby was soon to become another pioneer of flight. She moved to New York a few years later, and she took up flying (and writing about it) in 1911. She became enormously popular at the new air shows and, in 1912, was the first woman to fly the English Channel.[16] Ten weeks later Quimby did yet another air show back in America, and died when she fell from her cockpit.

We do not know whether Loughead (or her sons) and Quimby talked with one another, but it would make sense if they did. For Flora Loughead, mother of what became the Lockheed Company, was clearly driven by the same restlessness that has always been shared by pioneers, reformers, and inventors.

The forces of social reform, suffrage, and technology were running in overdrive at the turn of the twentieth century. They reflected a central belief in the power of radical change. That belief gave us an extraordinary explosion of change, and along the way it gave us the airplane.[17]

The antecedents of the Wright brothers, and the inventors who followed in their wake, produce countless stories such as these. This was a time filled with almosts, might-have-beens, and obsessions. Dreams were thwarted more often than they were fulfilled. Montgomery's claims were a great deal more plausible than most. Taken in context, they reveal how the general hunger for flight was being teased and driven in one place—in California. When that obsession finally touched the Loughead brothers—when they too were bitten by the bug of flight—Montgomery's efforts really took root.

So who invented the airplane? They all did, and none of them did. I've included peripheral figures along with more familiar aerial pioneers because the forgotten inventors reveal how the airplane was invented by a Zeitgeist, not by any one person. Together, this glorious company reveals the animating spirit that had to precede the grand successes.

If I am asked who gave us the airplane, I will check the box for the Wright brothers, and I will do so with absolute confidence. The totality of their success formed a great watershed between the dream and the actuality. But were George Cayley or Frederick Marriott less significant aerial pioneers than the Wrights? I do not think so. Montgomery was oh so close. Even off-the-wall Reverend Burrell Cannon was totally and seriously committed to the game. Flora Loughead never built an airplane, but she was surely a driving factor in the invention of human flight. You need only pick any of a thousand other Ötzis of aviation if you want to meet a true airplane inventor.

When I was young, I asked my father what he had heard about flight before the Wright brothers. He answered with the story of a local farmer in his Swiss/Austrian community of Nauvoo, Illinois. The man boasted loudly about the day he had somehow contrived to make a great leap off a barn roof. Had he used some sort of apparatus? My father did not know. He could tell me only how the man made the perfect butt of crude farm ridicule.

Yet this incident had touched my father; a scant fifteen years later, he joined the Army Air Service and learned to fly in a primitive Curtiss Jenny. Why did he gravitate back to that nutty farmer when I queried him? As I think back to his telling of that vague story, I think he was subconsciously trying to make me understand how flying was on the drum a century ago—and that we had reached the point at which we would no longer be denied.

Miller Aerostat, 1843.

Part II

Steam and Speed

4

Inventing Steam

"Alles was Odem hat"

The steam engine was every bit as civilization-altering as the airplane was. Yet it never could have embodied the animal insistence that the desire to fly did. The practical steam engine did not haunt people's dreams for millennia before they succeeded in creating it, the way flight did. If you or I have ever dreamt of steam engines, we have done so only on the basis of having first seen one.

What touched the human imagination instead were ectoplasmic wraiths—fog, vapor, air, wind, marsh gas. The antecedent concept was not the engine itself but the ghostly motive agent within it. The steam engine came into being only after we had finally extracted from our dreams a conception of vapor as capable of exerting rock-hard forces. It took thousands of years to do so. For that reason, we must undertake the pre-invention of the steam engine in two parts—first the steam, then the antecedents of the engine itself.

So how did we come to give steam its form and substance? If I were to be given the opportunity to talk with Ötzi, one of my many questions for him would be "Did you realize that air is a substance?" Actually, there would be more to that question. I would wonder, for example, if he realized what a powerful beast is unleashed when water turns to vapor. Did he suspect that air has significant weight?[1]

I refer this question back to Ötzi because I need a fresh way to look at it. You and I live with the motive power of gases and vapor every moment in our contemporary world. They propel our automobiles and illuminate our lightbulbs. They air-condition our sometimes feverish world. Yet we realize that we need some level of specialized knowledge before we can understand the material character of gases and vapors.

As early as Ötzi's time, one gas—namely, air—had already been harnessed for human use. Ötzi had probably seen, or even ridden in, a boat driven by sails. He also had some mechanical means for driving air into the charcoal fire with which he smelted copper. It may have been a bellows, a system of nozzles (or tuyeres) blown by human lungs, or some natural convection draft arrangement. He could not have reached a high enough temperature to melt copper without a strong air supply.

We cannot ask how modern internal and external combustion engines were invented without first trying to understand what it was to live in a world without all that you and I take for granted about gaseous behavior. I look to Ötzi because only by seeing through his eyes will we ever reconstruct any sense of how that modern knowledge emerged.

For you and me, steam speaks of power, force, energy—it speaks of both danger and useful potential. We think we see steam here, yet even *we* misread the signs. For this is only a cloud of water droplets that condensed, almost immediately, out of the vital but invisible steam leaving a locomotive.

So what *did* Ötzi know? He must, for example, at some time or other have spilled water on the heated stones that formed his smelting furnace. There had to be some visceral recognition of the power inchoate within the explosive hissing and white billows that formed as water first boiled and then condensed.

Did he equate those billows with the clouds above him? I expect that he recognized the kinship they held with the fog of early morning and the clouds in the sky. He was less likely to have realized that the billows above boiled water are water droplets, stripped of the gaseous energy they once held. He surely must, as we all have done at least once, put his hand into the clear vapor above a boiling pot—not yet condensed into benign droplets—and scalded himself.

Interpreting all that is no easy task. As we look closely at gases and vapors, surprises rise at every turn. Consider a contradictory difference between scalding steam and the air around us. Steam boils at 100°C (212°F). Like water, air also becomes liquid at a low enough temperature. At atmospheric pressure, however, that boiling point is incomprehensibly low—minus 194°C. That means the comfortable air around us is a gas superheated far above its boiling point, while scalding steam is (paradoxically) not superheated at all.

That much would have been quite beyond Ötzi's grasp, but he still must have digested a great deal from the clues around him. One thing we do know about Ötzi is that he not only recognized one gaseous form but also revered it. That form was air in motion. We know this not from the scant available evidence of late Stone Age life in northern Italy but by induction. It is something we learn by extrapolating facts about ancient languages.

All the old languages used a single word for wind, for breath, and for soul, and some of those words linger today. Go to the last line of the book of Psalms in an old German Bible, and you will read "Alles, was Odem hat, lobe dem Herrn." When a choir sings either of Bach's motets "Lobet dem Herrn" and "Singet den Herrn" in English, that line comes out translated as "All that hath *life and breath,* praise ye the Lord."

The obsolete word *Odem* does not just mean "breath." Rather, it means breath as the motion—the wind—of the human spirit. It embodies the belief that our spirit, our soul, rides upon our breath. In Sanskrit the equivalent word was *atman*; in Latin it was either *spiritus* or *anima*; in Hebrew today it is still *ruach*; in Greek it was *pneuma*; and in Chinese it was *qi* (or *chi*). Slavic languages use the word *duh* interchangeably for "breath" and "spirit." When we realize this, such

oddities as the Buddhist wind-driven prayer wheel make better sense to our secular, and far less metaphorical, thinking.

These words, along with their threefold metaphysical meanings, also reflect in phrases that we still use— phrases such as "not a breath

of air." The familiar expression "her dying breath" recalls the idea that one's spirit rejoins the wind, of which it was once a part. In the old view of breath/wind/spirit, one's spirit is, at that point, returning to God.

The metaphorical complexity of gases and vapors is wonderfully exposed in an old children's story about a traveler lost in a vast forest one winter's night. I have no idea how old this story is, or what metamorphoses it may have undergone. But I find it highly revealing.

The traveler stumbles into the hut of a mysterious woman—a woman of the forest. He begs for some warmth and sustenance. The woman says, "Yes, of course."

She invites him to sit by the fire, and she sets about to serve him a steaming bowl of soup. He blows on his hands while she ladles it from the pot.

"What are you doing?" she snaps.

"Why, my hands are cold. I'm warming them with my breath."

She eyes him suspiciously as she hands him the soup. He blows across the hot liquid in the spoon before he puts it in his mouth.

"Now what are you doing?" she cries, growing increasingly agitated.

He glances up, surprised, and says, politely, "The soup is so wonderfully hot. I simply meant to cool it before I try to swallow it."

At that point the woman seizes a stick of firewood and shouts, "Get out! Get out of my house! I'll have no sorcerer who can blow both hot and cold under my roof!"

Perhaps the purpose of the story was to tell children that they should be consistent. And that's something we all need to think about. But we learn something about consistency by looking at the *literal* act of blowing hot and cold. Air leaves our body at a little over 98°F. When we come in out of the cold, we open our mouth wide and exhale that warm air upon our hands. We clear our own air passage, so the warm air leaves almost unimpeded.

Cooling soup is another matter. Now we purse our lips to block the outflow of air and to increase the pressure of air in our lungs and mouth. That results in what is called an isentropic expansion of the air in our lungs, which we cause by creating this pressure drop across our lips.[2]

On either side of the expansion nozzle formed by our mouth, the ratio of the absolute temperature of exiting air to that in our body varies roughly as the one-quarter power of the ratio of atmospheric pressure to that in our lungs. So the air emerges cooler. How much cooler depends upon how hard we can blow. More than that, a relatively small pressure drop creates quite a fast-moving jet, and even a very small jet entrains a great deal of air. Thus what crosses the soup spoon is largely room air, drawn in by the jet. While our breath truly is cooler now, the soup is largely cooled by room air.

Pause and try it. Hold your hands up and blow upon them in both ways. There's no sorcery; we all really do blow hot and cold, and we do it instinctively. Ötzi did so 5,300 years ago, and even someone who knows all about air entrainment and isentropic expansions cools soup without giving conscious thought to the science of it. We need to keep in mind the pursed lips of that forest traveler, for we shall encounter them again in different forms.

Meanwhile, that story about the woman's reaction to the traveler remains—it and another one like it, about a king who grew frustrated with advisors who kept telling him, "On the other hand . . ." The king finally shouted to his chamberlain, "Go out and find me a one-armed advisor."

As we observe in Chapter 2, we all appear to carry an "anti-ambiguity gene." The intolerance for ambiguity is stronger in some than in others, but it is always present in some degree. At some point, we all desire the comfort of an unambiguous assignment of priority even if we struggle to avoid making it. Over the centuries, that anti-ambiguity gene has remorselessly stood in the way of our reaching the next level of understanding the true nature of gases and vapors.

Each time thermodynamic knowledge altered, it provided answers to many questions. Yet each such shift also left people clinging to other misapprehensions. When useful steam engines were finally built, that process accelerated. With increasing frequency, the engine forced people to revise what they thought they knew about gaseous behavior. (We like to tell ourselves that science drives technology. But, as we shall see, the steam engine was a great technological driver of science.)

At our best, we achieve changes in understanding not by avoiding blowing consistently hot or cold but by avoiding being maneuvered (by uncertainty) into blowing *neither* hot nor cold. Wright brothers biographer Tom Crouch tells how Orville and Wilbur would argue so ferociously that their friends had to listen closely to realize that they had exchanged positions during their verbal combat.[3]

Our dislike of uncertainty has a certain value, for only by embracing ideas fully can we really test and reject them (as the Wright brothers did). Perhaps, if we paraphrase Barry Goldwater, some extremism—some heat or cold—in the pursuit of truth may not be entirely a vice. The real vice is being unable to change when change is clearly called for.

The first step on the road to understanding steam was giving up the old idea that it is less than a real material. Gases had to be granted material substance. In one sense, the classical Greeks were already quite clear on that point. Western chemistry grew up around their ancient ideas of Earth, Air, Fire, and Water—the so-called Aristotelian elements. And yet, though those elements (or essences) combined to give us corporeal materiality, one might question the materiality of the elements, themselves.

The four elements had originated with the fifth-century BC Sicilian aristocrat Empedocles of Acragas.[4] Only parts of two book-length poems that he wrote survive, and only one of those dealt with nature. Empedocles is best known for asserting that all matter is formed when the opposing forces of Love and Hate act upon the elements.

That idea would not have done as well as it did had Empedocles not been a formidable figure in his own time. He appears to have written plays and treatises on medicine. Aristotle credited him with inventing rhetoric and with the idea that light moves at a finite speed. Empedocles talked about the interaction of matter and mechanical force, about attractive and repulsive forces, about mass and energy conservation. Freud wondered aloud if his own views of Love and Hate were subconsciously influenced by his early readings about Empedocles.

Empedocles even offered an early concept of natural selection. His biographer Alexander Mourelatos tells how he explained the origins of animals by "imagining random combinations of stray limbs and organs." Those combinations first produced only monsters, but viable arrangements eventually emerged and survived. Darwin quoted Aristotle's discussion of Empedocles's evolutionary ideas, then continued where Empedocles had left off. Empedocles's creatures, once randomly created, were either fit to survive or not. Darwinian natural selection went on to improve the ability of living things to function.

Naturally, an intellectual force as large as Empedocles acquired stories. In that, he resembled another figure of early history, Imhotep (whom we visited in Chapter 2). According to one legend, Empedocles was so convinced of his own immortality that he leapt into the crater of Mount Etna; in another, he simply ascended into the sky. Mourelatos tells how nineteenth-century Sicilian supporters of Garibaldi took Empedocles—who, like Garibaldi, was an aristocrat—as the prototype of their hero. Drawing upon many of the old legends, they fancied him as holding Garibaldi's populist sentiments as well.

When Aristotle got his hands on Empedocles's ideas about Earth, Air, Fire, and Water, he changed the external forces that acted upon the essences to transmute them. Aristotle replaced Love and Hate with physical counterparts—heat and cold or dampness and dryness. The evolution of chemistry was beginning.

This alchemy, as it came to be called, of Empedocles and Aristotle continued for millennia in guiding the struggle to understand gases and vapors. You and I therefore need to put aside our ingrained understanding of atoms and molecules if we hope to trace the way that understanding emerged.

And yet atoms had been suggested as early as Empedocles's essences. Our canonical inventor of an atomic theory is Leucippus, who promoted the idea around 430 BC. His student Democritus strongly advanced the idea. And, soon after Aristotle, the philosopher Epicurus was still promoting it—overshadowed though he may have been by Aristotle's acceptance of Empedocles.

Then, in the mid-first century BC, the Roman Epicurean philosopher Titus Lucretius Carus revived the idea in a powerful book-length poem expressing surprisingly accurate ideas about the nature of matter.[5] Lucretius left few tracks. We gain some clues about his life from contemporary Roman writers who mention him, and in the fourth century AD, St. Jerome wrote about him. But our best knowledge of Lucretius today is that one long poem, *De Rerum Natura—On the Nature of Things*. It expresses the Epicurean simplicity and lucidity of Lucretius's remarkable mind. He scorned the mystery-laden polytheism of Rome.

Epicureans believed that things are what they seem to be—that our senses do not deceive us.

Lucretius watched matter dividing, subdividing, and rejoining, and he concluded that matter must be made of tiny building blocks—of atoms that divide and rejoin. He failed to displace the already established Aristotelian essences, now on their way to dominating Western thinking until the seventeenth century. But his work survived because he was a superb poet. Latin was a practical and direct language, but Lucretius reshaped it and created beauty as he did. Listen to him as he paints a completely modern picture of atoms moving in solids and gases:

> . . . harried by tireless motion this way, that way,
> Some crash head-on and rebound over vast chasms,
> While some veer slightly apart from a close-dealt blow,
> And atoms that jar and rebound over tiny spaces,
> So tightly wedged in their assembly, tangled
> Up in their snagged shapes, intertwined and locked,
> These constitute the strong rock roots, the rugged
> Structure of iron and all things of like strength.
> But the rest, the few that wander the great void,
> Ricochet far and round from afar vast chasms
> In speedy return; these atoms furnish for us
> The thin air and the sunlight in all splendor.[6]

In that one passage we find intimations of atomic bonding, a full-blown kinetic theory of gases, even the notion of photons. And, when such nineteenth-century thermodynamicists as J. W. Gibbs talked about large sets of atoms, they called them "assemblies." That was pure Lucretius, for he had used the Latin word for the Greek citizens' assembly when he talked about aggregate atomic behavior.

De Rerum Natura was standard classical literature when modern science began taking its form in the 1600s. The poem was a rich trove of ideas—ideas out of place in time, modern ideas of heat, and Galilean ideas about falling—that would have been bypassed and forgotten had they not been so beautifully said. The ideas expressed in *De Rerum Natura* survived as poetry, but as science they fell by the wayside.

The early Catholic Church adopted a theology strongly based on Plato's philosophy, but Plato had offered little in the way of science. For their body of scientific knowledge, medieval Platonists went to Aristotle's scientific work. In it, they found a set of facts, rather than Aristotle's powerful strategy for expanding upon incomplete knowledge by the systematic use of observation. Until people once more recognized those methods and, using them, uncovered the mechanical character of the universe, atomism would be unable to prevail over the idea that matter was alchemical essence.

Atomism would have to wait for the seventeenth-century scientific revolution. Until we began working out the mechanics of particle interactions, gases would remain at least partially ephemeral. And only after air and steam gained corporeal substance would we, at last, create a steam-powered engine.

Makers of machines, however, were not idle in the meantime. During the three centuries that separated Aristotle and Lucretius, steam power almost *was* harnessed in the ancient world. A succession of Hellenistic engineers, based primarily in Alexandria, Egypt, recognized the material nature of air. They put to use its ability to exert large forces. Remarkable technological and practical scientific activity took place during that period.

Their contributions were one result of a major upheaval that took place between 334 and 323 BC. Alexander of Macedon had burst out of the narrow confines of Hellenic Greece and created an empire reaching across most of the Mediterranean and all of Asia Minor. When Alexander died young, his empire did not really survive him. What did survive were the effects of the cultural mixing that he had accomplished. In a very short time, classic Hellenic culture had mutated, spread, and turned into the astonishing Hellenistic world.[7]

A new cosmopolitan culture emerged throughout what briefly had been Alexander's empire. A version of the Greek language called *Koine* quickly evolved to serve as a lingua franca and helped to connect the administrative splinters of his empire.

This so-called Hellenistic culture was Romantic in character in much the same way as we understand that word. Hellenistic art and architecture were grand and theatrical, with little defining style and a strong taste for great works and emotional appeal. Greek, Oriental, and Arab themes intertwined. The Hellenistic era also produced a new literary form—the novel. Most important was the rise of a new appreciation for applied science. (All that is pretty much what happened again 2,100 years later, in nineteenth-century Europe and Great Britain. And, like the Victorian engineers, the ones in Alexandria also built for size and drama, doing so with vast imagination and intelligence.)

The Hellenic Greeks had used the wonderful word *techni* (τεχνη) for technique or skill in making either art or artifacts. In early Homeric Greece, such a user of *techni* as a sculptor or a mason was viewed with respect. During the golden age of Athens, Plato still said that *techni* was an innate human virtue, but he added that thinkers had enough to do without wasting their time with it. *Techni* should be the work of an underclass—of slaves and foreigners. Philosophy was business of a higher order.

That had to change in this new topsy-turvy, overblown Hellenistic world, so very different from classical Hellenic Greece. Alexandria was not even on the same continent as Athens. Into this stew pot of people, geography, and ideas was born, around 287 BC, a Greek who straddled the two worlds. The superb mathematician-mechanician Archimedes worked most of his life in Greece. Though he lived in Syracuse, he also spent time across the Mediterranean in Alexandria. Many of his surviving works were written in the form of his letters to Alexandrian colleagues.

Archimedes set the stage for a new interest in steam and air by beginning the shift into a more fruitful fusion of philosophy with *techni*. And yet he still talked the same way Plato had. Plutarch, for example, tells what happened when Archimedes so impressed his king with levers and pulleys that the king asked him to develop weapons. Archimedes said no. Then he added something Plato himself might have said: "The work of an engineer and every art that ministers to the needs of life [is] ignoble and vulgar."[8]

Putting aside questions about the extent to which war constitutes a ministry to the needs of life, Archimedes did keep his practical works under the table. Plutarch also says of Archimedes that he would not deign to leave behind him any writing on the subject of *techni*. Archimedes is known to us as a great inventor, but only through the huge tapestry of anecdote woven about him.

Perhaps Archimedes had in mind the words of Plato's contemporary, the orator Antiphon, who had said, "Mastered by nature, we o'ercome by art." For Antiphon, artisans were in the business of tricking nature. A lever or a pulley was a gadget that mocked the natural order of things. When Archimedes applied mathematics to machinery—when he mixed philosophy with base art—he knew he was crossing a line. He wanted to remain among the ranks of natural philosophers, not tricksters.

Two important engineers arrived on the heels of Archimedes, and they truly did step over the line at which Archimedes had halted—the line that separated philosophy and mechanics. Ktesibios (or Ctesibios) and Philon of Byzantium worked fully in the world of applied mechanics, and they led us back toward a real-world understanding of gases.[9]

The profoundly clever inventor and engineer Ktesibios was born in Alexandria in 270 BC. One story associated with him is that his father, a barber, contrived an adjustable mirror with a counterbalance that used a compressed-air plunger.

Whether or not that is true, it emphasized Ktesibios's comprehension of air's corporeal solidity. Ktesibios went on to create all kinds of air-powered machinery. He invented the earliest known water organ—a regular organ with a kind of water piston forcing the flow of air over its pipes. He invented a compressed-air pump and compressed-air catapult. (It is beyond our scope here to talk about his invention of the first feedback-controlled water clock. But it became the standard of timekeeping until the fourteenth century.)[10] The very name *Ktesibios* means "he who lives as a maker of things."

On the heels of Ktesibios, a Byzantine engineer named Philon appeared in Alexandria, and he created his own profusion of mechanical creations. He also left behind a book titled *Pneumatics*. In it, he described a dazzling variety of pneumatic devices, and they included many of Ktesibios's inventions.

A parade of Hellenistic steam and air devices followed. They continued until imperial Rome absorbed Egypt after the death of Anthony and Cleopatra in 30 BC. Even then, steam had yet to make its last and best-remembered appearance in Alexandria. Pick up any book on steam today—indeed, almost any book on invention—and it will mention Hero's turbine.

The turbine's creator was Lucretius's contemporary, Hero of Alexandria. He made very sophisticated machines during the first century BC, and his so-called turbine was a steam-jet-powered whirligig device that delivered no useable power. Nozzles (acting like the pursed lips of that forest traveler) on either side of a heated boiler drove its rotation.

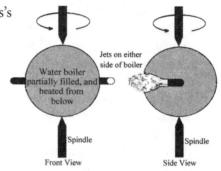

Approximate configuration of Hero's turbine.

By then, Alexandrian engineers had used steam and air to make jet-propelled birds that flew on the ends of strings, automatic door openers, and much more. Hero's turbine was only a late example of such machines. It marked the late days of this profoundly inventive age—this age that had struggled toward giving us a true steam engine long before such machines finally arose in the eighteenth century.

One other Hellenistic inventor had worked directly with steam, but she showed greater staying power in the long age of alchemy that covered the wake of Hero and Lucretius. She was a chemist called Maria the Jewess.[11] Maria has left fewer personal tracks than any of the others we have mentioned, despite her greater influence. Most of what we know about her comes from an Egyptian alchemist named Zosimos, who wrote in the late days of the Roman Empire, five hundred years after Maria lived. Among other things, Zosimos talks about her invention of a device called the *kerotakis*.

The *kerotakis* was one of many forms of stills, boilers, and reflux condensers that Maria invented. In it, she boiled mercury or sulfur in a lower container and used its condensing vapor to heat copper or lead in a pan above. It was a high-temperature version of the double boiler.

The familiar double boiler is an extremely clever contrivance. We cook food in an upper pan that is nested in a lower pan of boiling water. The food stays at the same temperature as the steam condensing below it—at 100°C. (The only reference to Maria that lingers in the modern world is the French word for a double boiler—*bain-marie*, literally "Maria's bath.")

Peerless Enameled Steel Seamless Milk or Rice Boiler.

No. 76392 Peerless Enameled Steel Seamless Milk or Rice Boiler. With retinned cover that fits both vessels.		
Quarts inside boiler	1	2
Holds, quarts	1¼	1¾
Weight, each, pounds	1¾	2
Each	41c	48c
Quarts inside boiler	3	4
Holds, quarts	3	4¼
Weight, each, pounds	2½	3¾
Each	60c	72c

A plain domestic double boiler, on sale in the 1900 Sears, Roebuck and Co. catalog.

Maria did much more than invent the double boiler, however. She founded an important school of chemistry in the late third century BC, and might even have met Archimedes in Alexandria when she was young. Scholars speculate on her origins. She's called "the Jewess" because Zosimos called her a "sister of Moses." That may have been no more than a convoluted way of saying she was wise. She could have been a Greek working in Egypt, or maybe a Syrian.

Maria was particularly interested in sulfur compounds. It was she who created means for making the silver sulfide that artists call niello. Niello is a matte black compound, often used for metalwork inlays. Her work was what we might call applied chemistry. Alchemists of a later age took a great interest in her experimental apparatus, and they used a sometimes mystical and metaphorical language to describe her processes; but she was closer in her thinking to Egyptian process engineers, like those who advanced the technology of brewing beer. Maria resembled today's chemical engineer more than she did the philosophers who went on to make such heavy use of her methods.

Alchemy took on many of its arcane trappings after the classical revival that attended the wildfire spread of printed books in the late fifteenth century. Maria's condensers, boilers, and stills were absorbed into an arsenal of chemical means (many of them aimed at transmuting matter) during the sixteenth century.

In any event, medieval philosophers had little further to say about the corporeal makeup of gases. Yet Lynn White Jr. describes several practical steam- and air-powered devices that arose in medieval Europe.[12] And these lay even further outside the academic world than the Hellenistic machines had.

One of the more common ones was the sufflator, whose purpose was to blow air into a fireplace. A sufflator was an andiron shaped like the body of some person or beast. It was hollow and partially filled with water. The head's pursed lips once more formed a nozzle. The fire heated the andiron and boiled the water. A steam jet, pointed at the fire, emerged from the lips (the nozzle) and entrained air into the fire, exactly as our woodsman entrained air to cool his soup. Similar steam jet ejector pumps are in common industrial use today, since a small flow of steam can entrain a large flow of air or other gas. Sufflators were effective machines, and, unlike the closely related Hero's turbine, examples survive today.

Fireplaces evoked the motive power of gases in many ways. White mentions another late medieval device—a small turbine mounted in a chimney that used the fire's updraft to drive a spit in the fireplace. That, in turn, immediately evokes the most significant gas-driven medieval device of them all, the windmill.

Windmills were kin to sailing ships but quite different in character from the jet-propelled Hero's turbine. In a windmill, the focus shifts from the action of a jet to something driven by a jet. That was a radical step forward. The windmills that appeared in Europe just before the year 1200 used fans facing into the wind. The turbine in the fireplace, however, mirrored the earlier Arab windmills, which were mounted inside closed towers. Louvers in their sides created a vortex of air within. They acted very much like a chimney with a turbine blade inside it.

As such uses of gases and vapors grew and evolved, they generally were not documented in writing. Nor did they yet serve as grist for the mills of philosophy. An academic understanding of the corporeal nature of gases would come from another quarter entirely. Kepler and Galileo set the stage for a renewed interest in atoms when they formulated a concept of celestial mechanics that could be applied to the movement of gaseous atoms as well as to planets. Only once we had mechanical means for describing atomic movement would the alchemical essences begin their slow fall from grace.

A key player in the rebirth of atomism was French philosopher Pierre Gassendi.[13] Born in 1592, he was twenty-eight years younger than Galileo. Gassendi became a priest and, between the ages of twenty-five and thirty, was given the job of teaching Aristotelian science. Since the noose of the Inquisition had not yet tightened around Galileo's neck, his earlier criticisms of Aristotelian mechanics were well known, and Gassendi had read them. Gassendi, already a skeptic, picked up his own attack. He had a particular edge in that he unabashedly put empirical flesh and blood on his philosophy.

Galileo, for example, had reasoned that if you drop an object within some moving framework, it shares in the motion of that frame. Gassendi accepted that but, knowing that it needed an experimental demonstration, arranged to have a weight dropped from the mast of a rapidly moving ship. The weight landed at the foot of the mast. If it had fallen in accordance with Aristotle's belief that the fall should be independent of the ship, it would have landed well behind the mast. In 1640, that experiment gave enormous validity to Copernican and Galilean dynamics. Indeed, it might have saved Galileo from some of his troubles had he himself taken the trouble to do it.

Gassendi did much more. He gave us a number for the speed of sound on a normal day at the earth's surface. It was accurate within 7 percent. But his greatest contribution was breathing life back into the atomic theory. Galileo had opposed the Scholastics' understanding of matter as alchemical essence, but Gassendi went further. First, he correctly said that all material phenomena arise from the indestructible motion of atoms—that the makeup of anything is based on incessant atomic motion. He used that fact to correctly explain air pressure.

A curious sidelight on Gassendi is that he influenced Cyrano de Bergerac, and Cyrano became an early science fiction writer.[14] The question of the efficacy of steam is reflected in Cyrano's book *Voyage to the Moon*. His hero made several attempts to get there, and in one he surrounded his body with glasses of morning dew. When the sun warmed the dew and drew it heavenward, it drew Cyrano's space traveler along with it.

In the early seventeenth century we thus began seeing gases and vapor with new eyes. The intensity of this buildup recently came home to me when I discovered, quite by accident, three successive names in the *M* volume of the *Dictionary of Scientific Biography*.[15] All were little-known seventeenth-century scientists; all were born in the last years of the sixteenth century, just about the same time as Gassendi. And the work of each was connected to work done by Galileo and Torricelli on gases, liquids, and vacuums. This was too much coincidence for a random throw of the dice, yet that is exactly what it was. My three *M* names were Magiotti, Magni, and Magnenus. They were from Rome, Milan, and eastern France, respectively.

Magiotti worked with barometers. He helped us see that water expands or contracts as it is heated or cooled, but that pressure hardly affects it. He was first to verify experimentally that the speed of water leaving a nozzle varies as the square root of the height of the water above the nozzle.

The second of the three, Magni (like Galileo), was eventually accused of heresy. That was partly because he had offended certain Jesuits, but also because he (again, like Galileo) had opposed the Church's science. Aristotle, he said, was wrong in saying that nature abhors a vacuum. To show that one *could* create a vacuum, he invented the barometer, independent of Torricelli and at about the same time that Torricelli did.

The third of our three *M*'s, Magnenus, specifically argued that matter was made of atoms, not Aristotelian Earth, Air, Fire, and Water. Magnenus began putting flesh and blood on atoms. He said, for example, that they had finite size and were not mere points. While Magiotti and Magni talked about barometers and vacuums, Magnenus told us what air was actually made of. He believed in only three kinds of atoms—atoms of earth, fire, and water. In that, his atomic theory was not quite as fully evolved as Gassendi's, but he did recognize that air (or any gas) was a thin collection of atoms of one kind or another.

Magiotti, Magni, and Magnenus are largely forgotten today—three more Ötzis of atomic theory, I suppose. Yet each displays the Zeitgeist that moved among the makers of seventeenth-century science. We would not have to look far to find a hundred more like them. Indeed, two of their contemporaries, Della Porta and de Caus, each had proposed means for pumping water out of closed containers by introducing pressurized steam into the containers. Della Porta did so as early as 1606.[16] While they said nothing about atoms, they directly put to use the corporeal force that gases could exert by virtue of their atomic solidity.

Della Porta and de Caus thus leapfrogged all the way to rudimentary forms of steam engines. Viewed in that light, the coincidence of finding Magiotti, Magni, and Magnenus together might be less impressive than it first seemed.

One person in this parade of contributors made explicit what all these before him had merely hinted. He was Otto von Guericke, an influential citizen of Magdeburg in Saxony.[17] Born in 1602 of a wealthy family, von Guericke was a diplomat and, in 1647, became the mayor of Magdeburg. That was also the year in which he became acquainted with, and fascinated by, Galileo's and Torricelli's work with air pressure.

He soon invented his own vacuum pump, and what he did with it was spectacular. By 1657 he was able to give the citizens of Magdeburg a remarkable demonstration of the hidden potential of the atmosphere. He carefully machined two hollow flanged hemispheres, twenty inches in diameter. When their almost perfectly smooth flanges were joined, they formed a single tightly sealed sphere.

He pumped the air out of the sphere so the force of the atmosphere alone now held the two halves together. Then he harnessed sixteen horses, eight on each side, and let them try to pull the halves apart. The horses could not do it; it would have taken a force of over two tons to separate them.

Von Guericke did other such demonstrations. While they were equally revealing, what lingered was the image of those sixteen horses vainly trying to separate two hemispheres held together by the force of airy emptiness. That was theater that could no longer be ignored. And it did not just stay in Magdeburg; in 1664, he took his sphere demonstration to the royal court in Berlin.

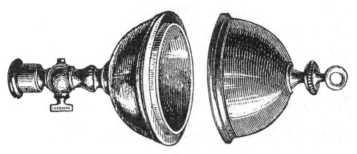

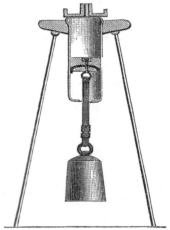

Above: The Magdeburg sphere.

Left: Another von Guericke experiment in which he lifted a heavy weight by drawing a vacuum in a cylinder into which was fitted a piston.

Von Guericke devoted the rest of his life to science, doing experiments in other areas and trying to evolve his own cosmology. But his legacy was his sphere experiment. He accomplished what all the others we have talked about did not. Once people had seen all those horses defeated by the enormous strength of seemingly fragile vapor, gases could never again be underestimated.

After von Guericke, it could only be a matter of time before someone would take the next step and create a work-producing machine—an engine—driven by steam.

5

From Steam to
Steam Engine

Upon the second-story ledge of the Napoleon Court of the Louvre are statues of the great French thinkers. Along with the likes of Descartes and Voltaire stands a man much less familiar to us, his left hand leaning upon a strange device. The statue is so far above the courtyard we squint to see what the device might be. It turns out to be a cylinder, cut away to expose a piston within it. Such a mechanism is mind-teasingly unexpected in this gallery of famous intellectuals.

The man holding it is Denis Papin.[1] He was born in Blois, France, in 1647—the same year that Otto von Guericke began working on vacuum pumps. Papin studied medicine and, by the age of twenty-two, had become a physician. Shortly thereafter, he went to Paris to work with the famed scientist Christian Huygens. Huygens had learned about air pumps when he visited the London laboratories of Robert Boyle in 1661. (Boyle, in turn, is best known for his work with gases and in particular for Boyle's law.)[2] Papin had made a vacuum pump under Huygens's direction, and he had applied it in experiments on the use of vacuums as a means for preserving foods.

In 1675, Papin made his own journey to London and was soon working as an assistant to Boyle. We need to consider what the word *assistantship* really meant. Steven Shapin, who has studied the relationship between Boyle and his assistants, quotes Boyle as having complained publicly about gentleman scientists who would not get their hands dirty in the lab.[3] He accused them of "effeminate squeamishness," yet he also placed his experiments in the hands of his many assistants. (The great scientist Robert Hooke had been among those assistants. He was working on Boyle's air pumps at the time Huygens visited.)

Boyle made little reference to his large laboratory support staff. Shapin quotes one mention of the "lusty and dexterous" fellows who operated his air pumps, and another instance in which Boyle tells an assistant to "proceed more warily" after the fellow almost killed himself doing a test. A particularly unsettling vapor of sanctimony wreathes our picture of Boyle when he says that it is a sign of Christian piety for scientists to be drudges and underlings in the search for God's truth.[4]

While he was with Boyle in 1679, Papin invented a device that he called a digester. (Its purpose was to "digest" bones for medical studies.) It was a very high-pressure pressure cooker—the prototype of the domestic ones that you or I might have used. Boyle actually mentioned Papin in the paper he presented describing the digester. Not only did he credit Papin with making inferences that Boyle did not need to check, he even credited him with having written the paper. That much concession was atypical for the times, and it actually bespoke an unusual fairness on Boyle's part. In the seventeenth century the right of publication came from the scientist's authority. No matter who ran the tests or even who wrote the words, what did matter was that the story was being told under his authority.

This touches our consideration of priority by revealing one more way we are led to replace the actual inventive process with canonical inventors. Even though we deem such "lab director" behavior as moral and professional anathema, the practice is not dead. Today one still finds scientists who in effect tell their assistants, "If you join my group, do my work, and write my papers, then my luminosity will shine upon you."

Three significant threads were interwoven in Huygens and Papin's lives. One was their association with Boyle. Another was the role that they both played in the evolution of practical steam engines. The third was that they were both French Protestants—called Huguenots.

The Huguenots had coexisted with French Catholics since 1598, when the Edict of Nantes had certified their freedom of religious practice in France. But in 1685, after a rapid buildup of anti-Protestant sentiment during the early 1680s, Louis XIV repealed the Edict of Nantes. As a result, four hundred thousand Huguenots were expelled from France. They had to move away into Germany, England, Canada, and other Protestant countries. Many Huguenots found their way to the British colonies in America. Paul Revere's father and Alexander Hamilton's grandmother were both Huguenots.

Huygens returned to his native Netherlands and never saw France again. Papin spent the remainder of his life in England and Germany. His peripatetic existence blunted his place in history—his place in priority. Both he and Huygens attempted to create an engine driven by gases, and neither was ever able to produce a working model. But Papin came close.

In 1673, while Papin was still working with Huygens in France, Huygens proposed an engine that he believed could be driven by gunpowder. Here again we see the probable mischief of scientific assistantships, for Papin later identified that idea as having been his own.[5]

By 1687, Papin had found a post at the University of Marburg, in what is now central Germany. Marburg had become home to many displaced Huguenots, and there Papin revisited the idea of a gunpowder engine. He soon saw that it would be impractical since it left so much "elastic" gas in the piston after each explosion.

Then it occurred to him that steam would condense to almost nothing and a piston stroke could be completed. Instead of exploding gunpowder to create pressure, he could condense steam to create a vacuum and let the exterior air pressure complete the working stroke. That, as we shall see, is how the first practical steam engines would eventually work.

In 1690, Papin published the design of a so-called atmospheric engine.[6] In his arrangement, the cylinder contained a small amount of water that was to be alternately boiled by an external heater, then allowed to cool. Condensation would be accomplished in the cooling, during which time the atmosphere would drive the piston downward. This engine is precisely the device that we see Papin holding today, on the ledge of the Louvre.

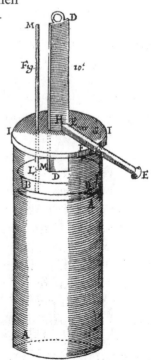

Retaining the water in the piston, as it turns out, slowed things down considerably. In the later practical engines, steam would be introduced from outside, then exhausted (either as spent steam or condensed water) through a system of valves. Papin reported that the heating part of the cycle took about one minute. Since cooling the cylinder was likely to take longer than heating it, one would be rash to hope that this device would give as many as thirty working cycles per hour.

Yet, despite its slowness, even this machine might have offered serious improvement over human labor in producing useful work. Papin would undoubtedly have continued to make improvements had his expatriate situation been better. His Marburg refuge was a difficult and low-paying post.

Papin's sketch of his first steam engine.

He finally managed to find a better position in Kassel, where he worked on a variety of pumps for a new patron. It was a position in which he had enough liberty to continue work on his engine. In 1707, he published the description of a much more straightforward high-pressure engine.[7]

Instead of using steam condensation to achieve the working stroke, Papin now proposed to admit high-pressure steam and let it do work by pressing the piston down. The spent steam would then simply be exhausted through a manual valve (G in the figure) into the atmosphere. Papin's friend Leibniz, in addition to being one of the creators of calculus, had a lively interest in a great range of technological developments. Papin wrote to Leibniz about this new engine, and Leibniz wrote back to suggest that the cylinder might be kept hot by running spent steam through a jacket around the cylinder. That, in fact, was a pretty sound idea.

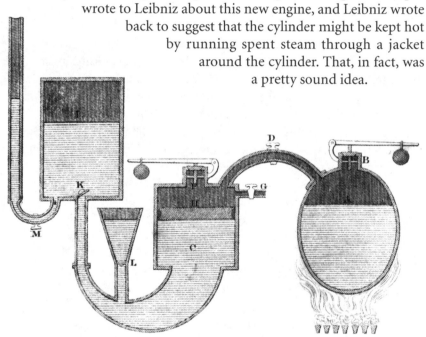

Papin's final steam engine design, as drawn by Lardner. Notice the pressure relief valves, which are adapted from his earlier digester.

Unfortunately Papin was stopped by the difficulties of manufacturing his engine, particularly of machining a piston and cylinder capable of holding high pressure. In the end, no Papin engine ever saw the light of day, and even as he worked in Germany, a Devonshire blacksmith named Thomas Newcomen was finally bringing a practical atmospheric engine into being—one that was kin to Papin's first engine.

But before we talk about Newcomen or any of the other successful engine builders, we should recognize that we have already encountered the conceptual form of every successful steam engine that would appear in the century following Papin. Let us therefore step back from specific inventions and weigh the essential problem for which any steam engine must be the solution.

Recognizing that steam can exert forces and thus promises to give us useful power, how do we employ that efficacy? If we create steam for such use, we must contain it; we must begin by boiling water in a closed container. That will almost certainly yield steam at a pressure equal to or higher than the atmosphere around it. Next, we need means for putting that steam to use. The history of steam power reveals that we have done so in two ways: using either a nozzle or a cylinder and piston.

First, the nozzle. When we release pressurized steam through a nozzle, it expands and forms a steam jet. The jet may act in either of two ways. It can act upon some sort of blading, like the medieval fireplace turbine mentioned in Chapter 4. However, working steam turbines came into general use only late in the nineteenth century, long after piston steam engines had become ubiquitous. Modern steam turbines use high-velocity steam jets, acting upon blades mounted around the rim of a rotating wheel.

Alternatively, the jet can exert a reaction force upon the nozzle, like the nozzles that drove Hero's turbine. Those steam jets differed from the jets that drive a Boeing 747 or propel a rocket into space only in that the gas was pure H_2O instead of some product of combustion. The boiler supplying steam to Hero's turbine operated at a pressure only slightly higher than atmospheric pressure, and its jets exerted only enough force to overcome air resistance and spindle friction.

Although nozzles were little used before the modern steam turbine, one highly evolved turbine appeared at the same time Papin was working with Huygens in Paris. The Jesuits, then a strong presence in China, were teaching, learning, and ordaining priests. In 1671, one of those priests, Min Ming-wo, built a set of steam-driven model boats and cars to entertain his emperor.[8]

Min's boilers supplied steam jets that impinged upon fast-turning paddle wheels. They in turn drove gear trains that slowed the rotation enough to drive a vehicle. Min Ming-wo created a complete functioning system in a model car that ran nonstop for two hours. Today, the Science Museum in Milan, Italy, has a working replica of it.

Yet Min merely put flesh and blood on a concept that had been published forty-two years earlier by Giovanni Branca.[9] Branca's book *Le Machine* included a clear and workable drawing of a steam turbine. In his drawing, the pursed lips of a sufflator (recall Chapter 4) drive a pair of pestles.

The second means for utilizing steam was the first to be brought to the point of practical use—applying steam to a piston within a cylinder.

That may be done in any of four ways, two of which Papin had already suggested: (1) Place a fixed amount of water in a cylinder, then alternately boil and condense it, as Papin did in his 1690 engine. (2) Boil the water in the cylinder and condense it outside. This would require the additional step of constantly recharging the cylinder with cold water. (3) Boil the water into steam outside the cylinder, then introduce it into the cylinder and condense it there. This is how the first working steam engines all operated. (4) Do both the boiling and condensing outside the cylinder.

Papin used the last of these means with his second engine, when he discharged spent steam into the atmosphere. That is also how a steam locomotive operates. Its *choo-choo* sound is the noise made by steam being exhausted into the open air where it then condenses. The birth of working engines would first revolve around method 3, and later appear in more sophisticated versions of method 4. The canonical inventor of the steam engine, James Watt, used method 4 exclusively and with many important improvements. But that would be much later.

We thus reached the beginning of the steam engine era with a vast amount of conceptual work completed, and a comparably large number of corporeal inventions had also prepared the way for the creation of working engines. Before we attempt to trace that transition, however, we should note that two driving forces conspired to bring an increased attention to bear upon the potential use of steam: during the late seventeenth century, the quest for economic and religious independence had developed an acute focus, and a serious energy crisis had appeared as well.

The pressure for liberalized religious practice was alive in England and France alike, and the Huguenot migration that had swept up Huygens and Papin helped to connect those movements in the two countries. I shall offer an obscure religious monograph to illustrate England's role in this rising religious tension. The title is *An Exposition of the Doctrine of the Church of England in the Several Articles* . . . (and it goes on).[10]

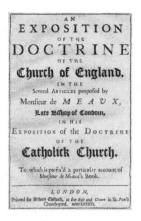

The author, a young cleric named William Wake, wrote it in 1686, the year after Papin arrived in London. This document gains significance when we learn that Wake later went on to become bishop of Lincoln Cathedral and then the archbishop of Canterbury. The ideas that he expressed would take on a curious importance as steam power came into its own in the early eighteenth century. Wake wrote it just after he returned from a three-year stint as chaplain to the English ambassador in France. He left Paris along with the Huguenots and then tried to delineate the limited actual differences between Anglican and Catholic doctrine. He chided the French for hardening their position and turning from reason to assault. Wake's tone is gentle and conciliatory, since he still hoped that reconciliation might be achievable.

Thirty years later, in 1716, Wake was made archbishop of Canterbury. From then until his death in 1737, he kept struggling to reconcile religious differences. As an old man he was found defending English-dissenting Protestants against his own Anglican Church. He tried to introduce prayer book reform that would reach out to and include the dissenters.

These British dissenters resembled people like Denis Papin in many respects, and they included Thomas Newcomen, who had just patented an atmospheric engine (of type 3, above). Newcomen was an ardent member of one of the emerging Baptist sects, living as Papin had lived, outside society's blessing. (Perhaps a new thing so radical as the steam engine had to come from those fringes of acceptability. Invention is, after all, by its nature a form of dissent and rebellion.)

The same people who were shaking off Wake's own English orthodoxy—the people to whom he now reached out—were the very ones who were forging a steam-powered economic revolution away from Anglican seats of power in London. The British Industrial Revolution was largely carried out in the English countryside, and the revolutionaries were tradesmen whose varying beliefs generally lay far to the left of the Anglican Church. They forged their freedom not by storming anyone's ramparts but by building a new industrial base for themselves (along with Great Britain) and by doing so in their own backyards.

The technologies that they took up included textile weaving, new kinds of iron production, and building a network of canals that decentralized the movement of goods. When Jacob Bronowski describes this period he does so under the title "The Drive for Power."[11] He talks about the other technologies, but he recognizes that power is the key.

The steam engine quickly took center stage among those technologies. In retrospect, one can see practical reasons why it did. The freedom being claimed by this new middle class could occur only as some of the physical burden was lifted from people's backs. The emergence of steam power preceded this countryside revolution—this tradesmen's revolution—and was then absorbed by the revolution. The original spur to the invention of steam power was a terrible energy crisis afflicting late-seventeenth-century Europe. A viable new engine was desperately needed.

It would be more accurate to call this the *second* great European energy crisis, for another one had occurred four hundred years earlier. The first crisis took place around the end of the twelfth century.[12] European millwrights had by then been building wooden waterwheels and windmills at a furious rate for around three hundred years. Then iron smelting had become an even worse destroyer of forests. To smelt one pound of iron took about eight cubic feet of wood, which had first been baked into charcoal.

As early as 1205, an Italian settlement had created its own reforestation plan in which each citizen had to plant ten trees a year, but that had been a gesture, not a solution. By 1230, the English had to start importing Scandinavian timber. The energy crisis of the twelfth century called for radical measures.

The English had been first to find a way around the problem. They found that they could replace wood fuel by burning surface coal, also called sea coal. Huge outcroppings of coal lay near the English coast around such places as Newcastle. When Marco Polo discovered the use of coal in China, in the late thirteenth century, he was unaware that the English were already using it for smithing, brewing, dyeing, and smelting, and that they were even exporting it to France.

Sea coal was filthy stuff, loaded with bitumen and sulfur. It created environmental problems from the start. Indeed, medieval environmentalists fought with medieval industrialists over the use of coal until famine and the Black Death tabled their argument. As Europe repopulated itself in the fifteenth century, coal had become the common source of heat. But now the developed techniques of metal mining were being used to get it. Miners could now follow coal seams deep into the earth.

They found their way down to the relatively clean hard coals that we use today, and such coal remained plenteous until the latter seventeenth century. Eventually, miners chased coal seams down to the water table. To mine beyond that level, they needed means for draining water from colliery shafts. Mine drainage required far more power than could reasonably be extracted from human laborers. Power was needed in regions without streams for water wheels. (And no wind could ever be sufficiently consistent to permit using a windmill to drain a mine with people in it.)

By the late seventeenth century, the dependence upon coal was huge, and the accessible supply was vanishing. What was needed was an engine to power the pumps. Indeed, if such an engine could be made to run on coal, the problem would be solved reflexively. But it was not yet apparent that an engine utilizing coal could be created.

So people did what they so commonly do under duress: they fed upon unreasonable hope. In this case the hope was that they might perfect a perpetual motion machine.[13] A perpetual motion machine produces usable energy without consuming energy in some other form.[14] Seventeenth-century scientists did not have a first law of thermodynamics to tell them that energy can never be created out of thin air. The concept—the hope—of perpetual motion therefore flourished.

As early as 1150, the Hindu mathematician Bhaskara proposed a machine to produce continuous power. His idea was simple enough: A wheel has weights mounted on arms all around its rim. The arms swing radially outward on the right and hang downward on the left, keeping the wheel forever out of balance. This so-called overcentered wheel should therefore turn forever.

In the latter seventeenth century, we still lacked the analytical mechanics that would show such a machine is impossible. No one yet had means for inventorying the various energies, kinetic and potential, involved in its rotation. While many suspected that the device was impossible, rational means for discrediting energy-creating machines were not yet in place. An overcentered wheel will start turning clockwise when it is released from the position shown in the picture, but then it hangs up. Arm A, which is poised to pass over the top, lacks the energy to flip over to the right and, when it fails to do so, the motion ceases.

I better understood the staying power of this device after I had the one in the picture built, and then showed it to knowledgeable technical people. Even when they were perfectly clear in their minds that perpetual motion is impossible, they reacted by suggesting ways in which I might improve the design proportions. My experience echoed the old story of the engineer who, stretched out upon a malfunctioning guillotine, could not resist pointing out to his captors how it could be repaired.

Author's overcentered wheel type of perpetual motion machine.

My little perpetual motion machine was profoundly seductive in much the same way.

After Bhaskara, the Moslems took an interest in the overcentered wheel around AD 1200. It showed up next in France in 1235.[15] For the next five hundred years, writers recommended the use of this ingenious (if impossible) device. And every time they tried to build one and failed, they concluded that they had the proportions wrong.

Eighteenth-century advances in mechanics finally made it absolutely clear that the overcentered wheel could never be made to work. But then, as new physical phenomena were described, people invented new means by which they believed each might be used to produce power without consuming energy. Perpetual motion machines based on static electricity, surface tension, magnetism, and hydrostatic forces are among the many that have been suggested.

Despite physical impossibility, inventors continually try to patent perpetual motion machines. The Patent Office says explicitly that it will not give a patent to any machine that violates the laws of thermodynamics. However, inventors often get around the legal law by obscuring their violation of the physical law. The office has frequently been fooled.

As the quest for a new energy source became more desperate during the last quarter of the seventeenth century, inventors proposed a host of variations upon the overcentered wheel. Reasonable people smelled a rat, of course, but they had no basis for refuting these ideas (other than the repeated failure of such machines).

All this came to a head in two important events. First, between 1686 and 1692, Leibniz developed the law of conservation of mechanical energy.[16] He made it clear that the kinetic and potential energies of any system (like the elements of the overcentered wheel and its surroundings) can be interchanged, but their sum will never increase. Thus the only way the overcentered wheel could ever give energy to the world around it would be to give up whatever minuscule kinetic or potential energy it held, and then come to a stop—something that invariably occurred within seconds.

The other critical event took place in 1698, only a few years after Leibniz's discovery. An inventor finally erected a working steam-powered pump and ran it where the public could see it. He was not Huygens, Papin, or Newcomen. He was certainly not James Watt, who would not even be born for another thirty-eight years. This inventor appeared in the mining country of southwest England, where the need for a power source was most severe. His name was Thomas Savery, and with him, talk of perpetual motion ceased for a season, while steam moved in to solve the problem of power production.

The road to Thomas Savery winds through the lives of three shadowy predecessors—or perhaps unpredecessors. A Scot named David Ramsay filed for a patent on a number of inventions in 1631. One was a device "to Raise Water from Lowe Pitts by Fire." We have no clues as to what the device was, but historian H. W. Dickenson is confident that Ramsay knew French landscape architect Salomon de Caus (Chapter 4), who proposed to lift water by squeezing it out of a tank with pressurized steam.[17]

Edward Somerset, soon to become the marquis of Worcester, knew Ramsay and went on to patent a so-called water-commanding engine. Cosmo de' Medici III visited Somerset's machine and noted that it could, by the work of one man, lift buckets of water a distance of forty feet at a rate of four per minute. The problem is that it takes only a tenth of a horsepower to do so. Human muscle was quite sufficient to the task. Was any steam involved? Subsequent writings by supporters of the marquis suggested that it was, but we have no hard evidence.[18]

Finally, the inventor Samuel Moreland, who worked for King Charles II, was credited with building a machine that raised any quantity of water "by the help of fire alone." But once more we have no hard evidence of the machine—no specific design.

The reason these people can be called Savery's predecessors is that they all dealt, or purportedly dealt, with the direct application of steam to lifting water. They enter our colloquy on priority in roughly the same way that some early airplane claimants entered it. Even if they represent more smoke than fire, they certainly do represent an amazingly active Zeitgeist.

Thomas Savery (c. 1650–1715) was an inventor, probably born in Devonshire, who worked in many technological arenas. He was interested in fortifications and navigation, among other pursuits, and he gained many patents. Most noteworthy of these was his 1698 patent for a "new Invention for Raiseing of Water and occasioning Motion to all Sorts of Mill Work by the Impellent Force of Fire which will be of great vse and Advantage for Drayning Mines, serveing Towns with Water, and for the Working of all Sorts of Mills where they have not the benefitt of Water nor constant Windes."[19] What Savery actually built fits into the story of the steam engine in an odd way. Although it is of type 3, his engine used, in place of a piston, only the surface of the water being pumped.

The heart of the engine was a tall, hollow pod. To describe its working cycle, we begin just as the pod has been filled with water. Steam, introduced into the top of the pod at eight or ten times atmospheric pressure, drives water out the bottom and upward. Then cold water is sprayed into the pod to condense the steam. The resulting vacuum draws water up from the sump below, and the cycle repeats. After his first engine, Savery went to two pods that could operate in alternation and produce a better

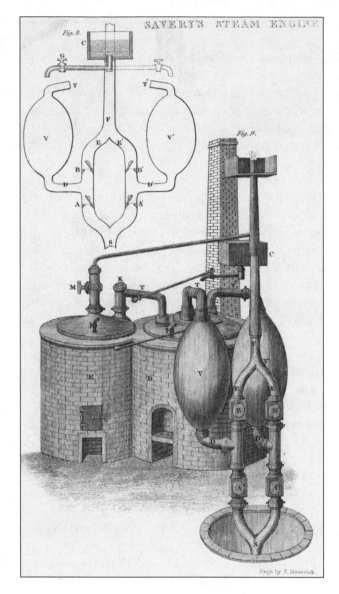

Savery's 1699 version of his steam pump, as shown by Lardner.

continuity of water upflow. He facilitated all this with a system of valves that he opened and closed manually in the right sequence.

None of Savery's patent claims about driving mills and other machines could have been achieved with this engine. It had no piston rod, no flywheel, no rocker arm, no means of any kind for driving machinery. His engine was a one-trick pony: it could only pump water.

Even as a mere pump, it had at least four serious limitations. First, the large pods posed a terrible pressure containment problem. Savery made them from soldered copper sheets reinforced by iron bands. The boiling temperature of water at his maximum intended boiler pressure of ten atmospheres is 357°F—sufficient to soften normal solder. He sought out special solders but was, even then, courting disaster.[20]

Second, Savery had no safety valve on either his boiler or the pods. Had he known of Papin's pressure cooker (the digester), he could have copied the weight valve that controlled its pressure. He did not, however, and that further aggravated the machine's susceptibility to explosions.

Third, the theoretical limit to the height that his pump could lift water from any sump below was only about thirty feet. In practice it could manage only about twenty feet. Such a low suction head meant that the whole installation could be used only near the bottom of a mineshaft, and that was no place to try to install a steam power plant.

The fourth problem actually alleviated the first two, but it also doomed Savery's pump for mine service. At ten atmospheres, the steam pressure should have been enough to push water another three hundred or so feet upward after it had been drawn up into the pods. The steam, however, condensed so rapidly when it made contact with the cold water that the pressure was reduced to less than one atmosphere above the surrounding air. Savery was lucky to push the water an additional twenty feet above the pod. Thus the actual net lifting capability was only around forty feet. Savery's engine never could have done service in a real mine.

At the same time, Savery did actually build his engine and did publicly demonstrate that steam could pump water. He also telegraphed his engine's potential by naming it "the Miners Friend." He left no doubt in anyone's mind that steam would be the answer to the desperately important problem of draining mines—even if his own engine was not up to the task.

Now our story's plot thickens because, while Savery was promoting his engine, another Devonshire man was also turning his attention to engine making. Thomas Newcomen was in business in Dartmouth with a partner, John Calley (sometimes spelled Cawley). They manufactured and sold goods made of iron, and Newcomen is often referred to as a blacksmith and an ironmonger. Calley is likewise often called a plumber. However, both were craftsmen whose abilities far outran such straightforward labels. Newcomen was well read and an activist in his church; both were ardent early Baptists.

The close proximity of Savery and Newcomen has teased historians for centuries. Did they know each other? Did Newcomen ever see a Savery engine in operation? Newcomen might even have made parts for Savery.[21] We know very little of this since Newcomen left no paper trail. We know almost nothing about the fourteen or so years during which he developed his engine.

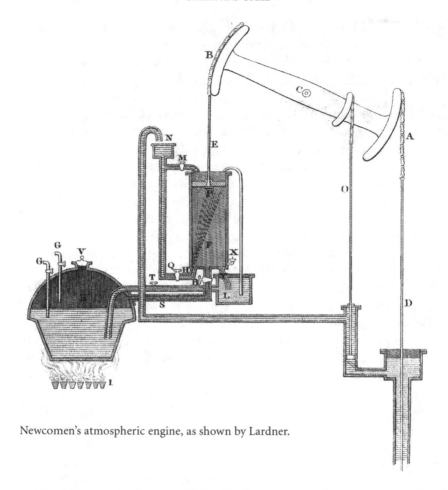

Newcomen's atmospheric engine, as shown by Lardner.

We do know that by 1712 he had built a steam engine radically different from Savery's. While both used external boilers and internal condensation, Newcomen's engines operated in a way that had a greater kinship to Papin's first engine, using a piston in a huge cylinder. In Newcomen's first engine, a cylinder almost two feet in diameter and almost eight feet high drove a large external pump. The system lifted about 130 English gallons per minute from a mine 153 feet deep. From that, we can calculate that the engine and pump together produced about 5.5 horsepower.[22]

But Newcomen's engine was not yet finished. In 1718, an inventor named Henry Beighton brought it close to ready for commercial service by automating the engine's valve action. Meanwhile, in 1716, a group of five businessmen had bought out Newcomen's interests (and Savery's as well). Thus, sometime after 1718, the engine became commercially viable. By 1725 Newcomen engines had found widespread use in England, and Europe was just beginning to put them to use as well.

Later-eighteenth-century Newcomen-type piston, roughly three feet in diameter. The wide circumferential recess accommodates a rope winding, which serves as a packing seal.

By 1763, Newcomen engines were so widely used that they had became a part of natural philosophy instruction at the University of Glasgow. And it is only at this late date that James Watt finally enters the picture.[23]

Watt was a Scot who, at nineteen, had spent a year in London studying instrument making. He had returned to Glasgow to open his own shop and within a year was given the title "Mathematical Instrument Maker to the University." He worked on many things during this time (among them, oddly enough, was making musical instruments).

Then, in 1763, a professor presented Watt with the task that changed history. The Newcomen engine model, bought as a demonstrator for that natural philosophy course at the university, had never run for more than a few strokes. It had even been sent off to a distinguished instrument maker in London, who had failed to make it run.

What was wrong with it? Could Watt identify and eliminate the problem?

Watt saw the problem, but to find a solution was another matter. Each time steam was admitted, it entered a cylinder cooled by the condensation stroke that had just taken place. Most of the steam that entered the cylinder condensed on its cold wall. Very little of it remained to create a working vacuum when it, in turn, was condensed. This same problem had to be present in full scale Newcomen engines; it was just made worse by the small size of the model. To improve the model would be to improve the engines themselves. It would require devising a way to keep the cylinder walls hot after the steam was condensed.

In 1764 Watt finally saw the way around the heating/cooling conundrum: if he could separate the condensation from the cylinder, he would no longer have to worry about the cylinder temperature. The cylinder

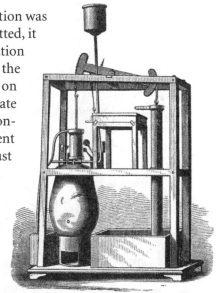

The Glasgow model of a Newcomen engine presented to Watt for rework.

would simply stay hot and the condenser cold. No more wasted steam! That solution, however, was less simple than it might seem. We have noted that Denis Papin exhausted the steam from his second engine and allowed it to condense in the open air outside the cylinder. But that was a high-pressure engine.

Watt, on the other hand, was trying to improve the existing Newcomen atmospheric engine—one in which condensation had to occur in a vacuum. Watt's external condenser would have to be a very low-pressure container. To accomplish that, he exhausted the steam into an external cylinder held at low pressure by a vacuum pump. If he could manage that, the cylinder would remain cool while he condensed steam at a pressure far below the surrounding atmosphere.

This was a huge inventive leap, and it was part of what made 1764 a very significant year for Watt. He also married his cousin Margaret Miller that summer. Everything appeared to be coming up roses for James Watt.

Now he had to make a working model of his new engine, and that was another matter. Soon he found that he was going broke, so he sold two-thirds of his patent to an inventor and speculator named John Roebuck. Roebuck's iron works were in trouble because its coal mines were flooding. He hoped to use Watt's engine to pump them dry, but the engine still had too many bugs to do the job. Then Roebuck made some bad business gambles, and by 1769 he too faced bankruptcy. He had to sell off his holdings to get out of debt, and no one would put up a farthing for his share of Watt's patent.

Finally the brilliant Birmingham industrialist Matthew Boulton saw what others did not. He gave Roebuck £1,200 for his share of the patent and thus bailed out both Watt and Roebuck. Watt kept soldiering on. He wrestled with his engine for the next four years while his wife tried to buoy him. "If it will not do, something else will," she wrote. "Never despair." Meanwhile, two of their children died as infants. Then, to make trouble complete, Margaret died in 1773, and the last threads of Watt's life seem to have come undone. He wrote, "I know grief has its period; but I have much to suffer."[24]

Yet his fortunes were about to turn once more. He and Boulton now held the patent for the external condenser. By 1776 his engine was running, and he moved to Birmingham to go into production with Boulton. That was the year he remarried. It was also the year when Samuel Johnson's biographer, James Boswell, visited Boulton's works. He writes of the experience, "I shall never forget Mr. Bolton's [sic] expression to me: 'I sell here, Sir, what all the world desires to have—POWER.'"[25]

That was surely one of the great double entendres of all time. In it we read the fulfillment of the path that the English dissidents had chosen.

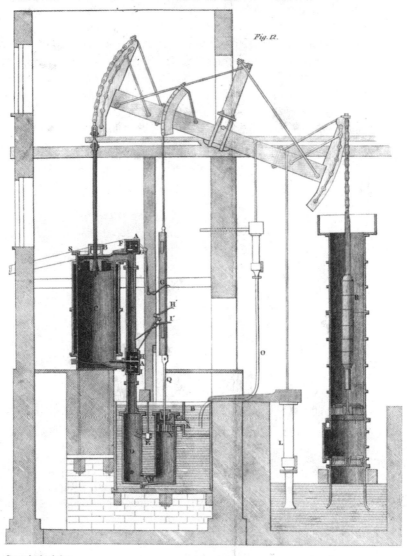

WATT'S SINGLE ACTING STEAM ENGINE.

Watt's first engine design (after Lardner).

They had embraced steam power at the turn of Watt's century; now they had economic power. Watt's external condenser had roughly doubled the efficiency of Newcomen's atmospheric engine, and that was only the beginning of his accomplishments. By 1784, his engines had (among other improvements) these very important energy-saving and weight-reducing features:

- *Straight-line mechanism.* A linkage of rods just below the rocker arm caused the upper end of the piston rod to move in an almost perfectly straight up-and-down path, while the rocker arm itself traced an arc. Watt also used a form of pantograph mechanism to copy this motion into the upper end of the vacuum pump piston rod. Watt's outwardly simple but highly sophisticated arrangement greatly reduced friction and wear on both pistons. (A pantograph allows one to trace a drawing with a stylus while a pen or pencil replicates it at a larger or smaller scale. Invented in 1630, it was very popular in the eighteenth century. Thomas Jefferson used a pantograph mechanism of his own design to make copies of the letters he wrote.)

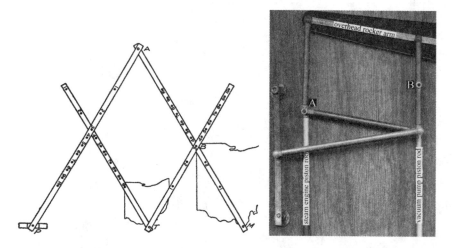

Left: A pantograph device tracing a small outline of the state of Ohio at point T, and outputting a large image at point M. *Right*: A model of Watt's straight-line mechanism showing how rods are arranged to produce an almost perfectly vertical motion of the engine piston at point A. This motion is pantographed into point B, where the vacuum pump piston rod attaches from behind. This model was built by Simon Dorton as an exercise in his history of technology course with the author.

- *Double-acting cylinder.* Watt created a symmetrical double-acting cylinder in which steam drove the piston on one side while condensation drew it on the other. He now had two steam engines for the price of one.
- *Early cutoff.* Watt created a valve arrangement that began exhausting the steam before the stroke was complete. That way he admitted roughly a third as much steam as he had previously done, but that steam continued doing work while its pressure decreased, after the exhaust valve was opened.

- *Flyball governor.* Watt invented a governor whose weighted arms opened outward as the engine sped up, and dropped as it slowed. That movement in turn drove a mechanism that changed the amount of steam admitted to the engine. Thus when the load on the engine was reduced, the steam supply to the engine would also be reduced, and the engine would continue running at the same uniform speed. As a result, less steam was consumed, and the engine's efficiency was again increased.

Despite all that Watt did, the steam engine had been "invented" long before he arrived, and who would be so rash as to name its inventor? If we really feel the need of naming a canonical inventor, that palm might go to Newcomen. Yet we can only stand in awe of all Watt achieved. He brought unparalleled inventive brilliance and energy to the task of consolidating and vastly improving a technology that already had a long history.

Now that we have the steam engine, we are summoned to another question about the course of invention. Notice that we have met one person

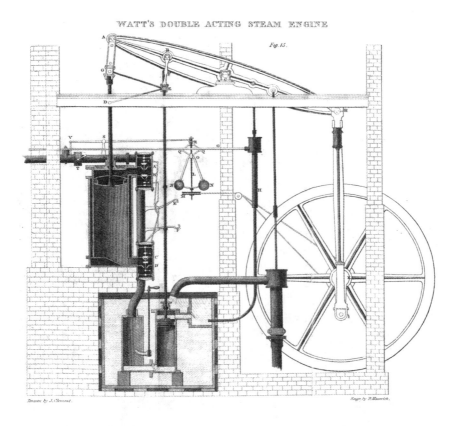

A more fully evolved Watt engine (after Lardner).

after another who helped to bring the engine into being, and all of them worked without access to a science of heat. The natural philosophers around Watt were still speaking of phlogiston and caloric, even of the original Aristotelian essences. We now had the engine without a science to describe it.

But the engine cried out to be explained. We have hitherto asked who invented this or that—the doughnut, the airplane, the steam engine. We have not yet asked who invented a science. The science of *thermodynamics* now had to come into being. And its inventor? Well, I shall suggest that its inventor was the steam engine itself, more than any person. Let us play next with that strange claim.

6

From Steam Engine to
Thermodynamics

Thermodynamics: the study of transformations of energy and how such transformations affect matter.

By 1800 we had a pretty clear sense, albeit a Rube Goldberg one, of how our new steam engines worked. We knew how to boil water under pressure. We had created ways to make the resulting steam produce a force upon a piston. We had many means for utilizing the forceful movement of a piston rod.

No doubt that felt like understanding, but there was a catch. A festering question remained—you might even call it a canonical question, for there it lay in the shadows of all we thought we knew. The question was: What does coal contain that can be transmuted into useful work? That steam-engine-generated question led to the invention of thermodynamics. Like any of the other of the human creations we have been discussing, thermodynamics has its canonical inventors—largely nineteenth-century thinkers with such names as Carnot, Joule, Kelvin, and Clausius. Yet thermodynamics also has its own long, winding history followed by a late-converging Zeitgeist. It has its many Ötzis. What makes the invention of thermodynamics unique is the steady chuffing sound of steam, ever present in its background.

Thermodynamics holds an odd place among the sciences. You might even want to regard it not as a science at all, but rather as a set of conclusions obtainable from atomic physics. And so it might seem. However, for reasons that still perplex us, we cannot quite derive all of thermodynamics from what we know of atoms. It retains life of its own as an independent science.

Like Euclidean geometry, thermodynamics is an arrangement of logical deductions resting upon a few empirical laws. Those laws may be framed in different ways. However, two primal facts underlie any set of thermodynamic axioms. One is that energy is conserved. The other is that the universe constantly moves toward more probable (and less orderly) states.

Each of these two facts of life severely constrains the working of any power-producing engine.[1]

The first, energy conservation, is patently obvious when we reduce it to summing the many ways in which energy is stored and traded on the atomic level. However, it is not so easy to derive the tendency toward disorder from atomic behavior. To this day, it still requires some logical arm waving to do so.

Even if we do eventually learn to deduce thermodynamics completely from atomic physics, we should remember that it came into being while atomic descriptions of matter were in their infancy. *It is a science of heat that can be expressed without reference to atoms.* While its physical axioms are necessarily rooted in atomic behavior, they express general truths that apply to vast aggregates of atoms—aggregates large enough to be experienced by human senses. To make calculations about real systems on the human scale, we will always have to turn to thermodynamics, even when our understanding of atomic physics has become far better than it presently is.

We noticed in Chapter 4 that certain thermodynamic concepts originated in antiquity. Those ancient concepts stayed fairly static until seventeenth-century thinkers began asking pointed questions about the substance of gases and vapors. That, in turn, reanimated a general curiosity about the nature of heat. From then on, we began revising those ancient notions.

It might help to pick up that thread with one particularly odd character, one person among the many who were pushing us toward a new understanding of heat. He was Johann Joachim Becher, born in Germany in 1635.[2] Becher edited an alchemical tract when he was only nineteen. From then on, his life seemed to revolve around the twin subjects of alchemy and gold.

He published his first book when he was twenty-six. It established him as a metallurgical chemist. But he lived a restless life. He worked for a while in Munich as an advisor to the elector of Bavaria, then in Vienna as commercial counselor to the Austrian emperor. During that time, he proposed building a canal connecting the Rhine and Danube Rivers to facilitate trade with the Dutch. (Such a canal, connecting the Rhine to the Danube through the Main River was finally completed in 1992.)

Becher also argued that governments should strictly control the flow of goods and money—that colonies should make up deficits with raw materials, chiefly with gold. His 1668 book, *Political Discourse,* established him as the leading German theorist of the economic system that we came to call mercantilism. After ten years in Vienna, his radical ideas got him fired. He was even jailed when he protested the importation of French goods.

So Becher moved to Holland. There he tried to sell the Dutch assembly on a process for extracting gold from sea sand. He built a small demonstration process, then suddenly decamped—he deserted his family and ran off to England without building the pilot plant. His last book was a chemistry text published in 1682, the same year he died. Its 1,500 chemical processes included one for making a "philosopher's stone," the purpose of which is to turn lead into gold.

Two years after Becher's death, Newton wrote his *Principia*. The rules of natural philosophy were changing in such a way as to leave Becher behind. Europe embraced his mercantilism, and almost immediately began pushing it to various breaking points. Mercantilism said that countries should specify their trade balances in advance. Then they should use raw materials from colonies to supply industries at home in such a way as to maintain those balances.

Spain's interpretation of mercantilism had been simply to steal vast amounts of gold from her Central American colonies and spend it without building up internal industries. She was already losing her standing as a major European nation by the time steam engines were coming into use. The mercantile economy in France was being controlled by such a strong centralized system that it would finally incur violent revolution.

Although Great Britain's American colonies eventually revolted against its mercantile practices and broke free, Britain's internal controls were not as stringent as those in France. Hence it was possible for a corrective industrial revolution to take place within Great Britain without major political upheavals and violence.

But Becher did not live to worry about any of that. He had spent his life skipping from stone to stone on the rivers of turmoil. At the end, he made a widely quoted remark that is probably better remembered than he himself is—a remark that does much to explain his strange life:

> Chemists are a strange class of mortals, impelled by an almost insane impulse to seek their pleasure among smoke and vapor, soot and flame, poisons and poverty, yet among all these evils I seem to live so sweetly, that [I'd die before I'd] change places with the Persian King.

While Becher appears to have ridden many of the wrong horses, things are not always as they seem. His scientific legacy proved to be quite large because he advocated an important shift in our view of the four Aristotelian essences. First, he renamed them "earths" (a term that lingers today in the group of metallic elements called the rare earths).[3] What had previously been the first essence, Earth, became "salty earth." Water became "mercurial earth," and Fire became "sulfuric earth." As for Air, he adopted the idea that it was inert and did not enter into any alchemical process. (Another century would pass before we knew about oxygen.)

Becher had an important follower, Georg Ernst Stahl, who completed his line of thinking.[4] Stahl became a giant of early-eighteenth-century science. He was born in Germany around 1660, became a doctor, and spent his life as an academic working in chemistry, medicine, and philosophy. He is sometimes described as harsh and intolerant, but biographer Lester King thinks the evidence for that is flimsy. Stahl's personal motto tells us as much as anything can about his personality. He expressed a powerful distaste for majority thinking when he said, "Where there is doubt, whatever the greatest mass of opinion maintains . . . is wrong." As the new experimental forms of science gained footing, along with the rising conviction that matter is made of atoms, Stahl fit into the ongoing scientific revolution in an odd way.

He was intensely religious, and he believed in a living anima, or soul. But his anima was a real force that impelled all motion within the body, including blood flow and all the other movement of bodily fluids. He disliked the way doctors were starting to analyze the body in terms of component parts and processes. How could you treat it piecemeal— dismantle it like a watch? The anima, after all, inhabited the *whole* body. (As we look at the whole-body trend of medicine today, we might wonder whether that places him far behind his time or far ahead.)

Stahl's belief in atoms is equally perplexing. He adopted Becher's view of the three earths, but he believed that they, and air, were made of constantly moving atoms. He was particularly interested in the substance of heat, and in 1697 he provided a very important reworking of its description.

The Greek words for "fire," φλογο (phlogo), and "fiery," φλογισοζ (phlogistos), had already been applied to the Aristotelian essence of Fire. Stahl renamed sulfur earth as "phlogiston" and gave it a new interpretation. His phlogiston combined with, or dissociated from, matter during any chemical reaction. It could also be stored in or taken from the atmosphere. When I incline to view phlogiston as a primitive belief, I have to remind myself that it is not so different from our talk about the heating values of fuels, or heats of reaction.

In Stahl's view, phlogiston was the element released when combustion takes place. He would say, for example, that carbon is a combination of ash and phlogiston. He regarded pure carbon as condensed phlogiston, which, after combustion, was released into the air in gaseous form, and finally regenerated in plants. Those views contain remarkable elements

Wax turning into phlogiston and soot.

of our present understanding of how combustion and photosynthesis work, and they remind us of the way in which the heat of reaction (bound up in molecular structure) is absorbed or released when any chemical reaction takes place.

By the time Stahl was writing, Hooke, Boyle, Leibniz, Descartes, and even Francis Bacon before them had attacked alchemy as it was then understood and practiced. Each had promoted the notion that heat is a manifestation of atomic movement. But (and here we are back to the very existence of thermodynamics) atomic theory offers cold comfort when we need to make a calculation. Let us imagine that we need to explain the workings of a steam engine in strictly atomic terms to a layperson today. I shall attempt to do so, step by step.

- Every molecule is held together by powerful electronic forces. It requires energy to pull those atoms apart. If we burn carbon in oxygen, we dismantle the O_2 molecules to form two atoms of oxygen. Then those two atoms combine with one carbon atom to form carbon dioxide, CO_2. Since it takes more energy to pull CO_2 apart than it does O_2, the net effect of burning carbon in oxygen is to *release* energy.
- That energy release increases the random speed of the newly created CO_2 molecules. Those fast-moving molecules impact the metal boiler. They agitate the atoms of metal and speed up their rate of vibration. The moving atoms of metal, in turn, agitate the water molecules inside the boiler. Their temperature increases until the water reaches its boiling point.
- The boiling point is the temperature at which water molecules become so agitated that they can coexist either as close-packed liquid molecules that merely jiggle or as highly separated, rapidly moving gas molecules. (It takes a great deal of energy to move all the water molecules from the liquid state to the gaseous state at its boiling temperature.)
- Those highly energetic gaseous H_2O (steam) molecules impinge upon a piston, acting like a vast number of tiny ping-pong balls, and their aggregate effect is a large and seemingly steady force. Under that force the piston moves, doing useful work in some way that we have contrived for it to do, depending upon the machinery we choose to build around it.

Even today, if we design power-producing machinery, we avoid trying to describe such atomic/molecular scenarios. They are little help when we set about to do calculations. For calculation, we need to use physical laws that no more recognize atoms and molecules than our own physical senses do. Not until the mid-nineteenth century would a detailed atomic/molecular understanding of nature begin to gel, and when it did, its purpose would *not* be to design engines.

The idea of phlogiston therefore survived long after Stahl. Throughout most of the eighteenth century it served philosophers and engineers alike as they continued to use it in their interpretation of heat phenomena. And as they did, they were able to expose another conceptual hurdle.

In hindsight we might wonder why one key difficulty with the phlogiston theory was not evident from the outset. Since phlogiston was a component of matter, it could be manifest only when a chemical transmutation takes place—such as producing ash and phlogiston from wood by igniting it. What then happens when, for example, we put a heated brick into a bucket of cold water? Despite the fact that no transmutation of matter takes place, the brick nevertheless cools and the water warms.

By the late eighteenth century two scientists in particular were picking a path through this difficulty by asking the right questions. They were Antoine Lavoisier, in Paris, and the chemist Joseph Black, who worked in Glasgow and Edinburgh.

Black was a Scot who happened to have been born in France, in 1728.[5] He did his undergraduate work at the University of Glasgow, then received an M.D. at Edinburgh with a dissertation on the use of magnesia as a stomach antacid. In 1756 he was made professor of chemistry at Glasgow, and he later moved back to the University of Edinburgh. From then until his death in 1799, he made major contributions to the task of sorting out the lingering mysteries of heat.

Phlogiston was still the prevailing conception of heat when Black was young. For that reason, Black first viewed heat as wedded to chemical change. As he began his inquiries, he concentrated on the *quantity* of phlogiston (or whatever the matter of "heat" might really be) that had to be added to a material to increase its temperature. He recognized that the amount of heat/phlogiston/energy needed to raise the temperature of an object one degree depends upon what the object is made of.

By then, we had excellent means for measuring the *intensity* of heat, or temperature. Thermometry was a full century and a half old when he began his work on heat. Both Fahrenheit and Celsius had long since honed the development of temperature scales. But how do we measure the *extent* of heat? Black realized that he could heat (or cool) a mass of water by transferring energy to (or from) it from (or to) a mass of some other material. Since the heat leaving one mass is the same as that entering the other, he could determine the heat capacity of any material relative to that of water.

He also took an interest in the kindred concept of latent heat. When a liquid is boiled or condensed, or a solid is melted or frozen, the transmutation takes place with no change in temperature. That is why the heating required to accomplish the change is called *latent*. Latent heats could be measured in pretty much the same way as specific heats. One might, for example, surround a known mass of ice with a known mass of hot water, and then measure how much the temperature of the hot water drops as the ice melts away.

In doing such tests, Black was setting the foundation for the concept of a British thermal unit, or Btu (a Btu is the energy required to cause a one degree Fahrenheit increase in the temperature of a pound of water under certain standard conditions). The Btu was the inevitable consequence of relating all these observations to changes in a pound of water.

Two examples of thermal property observations from an 1866 text.

Left: Thermometer immersed in liquid sulfur. When the fire is turned off, the temperature drops until the sulfur begins to freeze. No further temperature drop occurs until the sulfur is solid.

Right: Papin's digester (Chapter 5). Originally conceived for leaching marrow from bones, the illustration shows it being used to observe boiling and condensing at high pressures. (Note the pressure cooker type of control valve on top.)

Black at first regarded these changes as reflecting chemical changes in matter, but his view shifted. And here the steam engine surfaces once more, for James Watt had also tripped onto the idea of latent heat in his work. Watt was no phlogiston adherent; he simply faced the problem of keeping the walls of a Newcomen engine hot during the condensation portion of the power cycle. He realized that condensation caused so much cooling of the cylinder walls because water absorbed a latent heat—a very *high* latent heat. At that time, Watt and Black had become friends at the University of Glasgow. They compared ideas, and Black was able to confirm the concept for Watt.

Thus it was becoming clear, all around Black, that heat was not some component of matter, as phlogiston was presumed to be. Rather, it was separate from matter. It flowed in and out of matter. Another term, coming into general use, was about to displace phlogiston. That new term was *caloric,* and it would gain its full definition in the late 1770s, when Black supervised the dissertation of a bright young man named William Cleghorn.

Cleghorn had been born in 1751. Since his father had died young, his uncle George Cleghorn raised him and his eight siblings.[6] Uncle George was a noted physician at the University of Dublin. He had once served with the British army on the Mediterranean island of Minorca, and his book on the island's epidemiology was still in use in the nineteenth century.

Cleghorn finished his M.D. degree in 1779 and in his dissertation created a systematic description of caloric.[7] Black's lecture notes, which were published after his death by another former student, John Robison, include Cleghorn's rules of caloric.[8] Those rules were important because, while they helped to make a useful tool of caloric, they also helped to expose its eventual failings. They were:

- *Caloric is a subtle invisible fluid, which can be neither created nor destroyed.* (That captured the essence of energy conservation but failed to recognize that work can be dissipated into heat, as when we rub our hands together to warm them.)
- *Caloric is elastic, with particles that repel one another.* (That explained thermal expansion.)
- *Particles of caloric are attracted by particles of ordinary matter in different degrees for different substances and different states of aggregation.* (That accounted for differences in specific heats of substances. Cleghorn's use of the term *particles* reflects the increasing acceptance of atomism by then.)
- *Caloric is either sensible or latent.* Latent caloric is caloric that has combined chemically with particles of solid to form liquid, liquid to form vapor, or solid to form vapor.
- *Caloric has weight.* (This accounted for the increase of weight of metals when they were heated in the presence of air.)

The last rule came about because Cleghorn did not yet know about oxygen and oxidation. At the same time he was working on caloric, three other people were simultaneously working on the nature of combustion and coming to an awareness of the role of oxygen.

In 1774, the dissenting minister Joseph Priestley had isolated oxygen but misidentified it. At first he thought he had isolated nitrous oxide. Then he decided it must be air with the phlogiston removed. Antoine Lavoisier read Priestley's work as he too studied the problem. He finally identified the substance as a new element, in 1777. Meantime, Swedish apothecary C. W. Scheele had independently identified oxygen before either Priestley or Lavoisier. Scheele might have been called its discoverer had he published his results sooner than he did.[9]

On the matter of caloric, Black himself remained cautious. He knew that Cleghorn's rules were not the whole story, but he allowed that they explained experiments reported by Benjamin Franklin and others. Lavoisier was fairly happy with the idea, for it was he who had named it caloric (actually *calorique*). And Black, in the end, said of the caloric theory that it was "the most probable of any that I know."

So Cleghorn's caloric remained in use for the next seventy years. Not until atoms were far better understood would it become clear that heat is merely the way we perceive our body's overall response to atomic motion. However, the concept of caloric still serves us, even today. We speak of heat "flow," and about bodies "holding" their heat. To deal with heat in everyday life we still conceptualize it as behaving like a fluid. Learning, at last, that it is actually a manifestation of atomic motion did not change the way we visualized it.

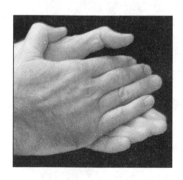

Creating caloric.

Yet Ötzi rubbed his hands together to warm them, as his forebears did from the time they'd first walked on two feet. Ötzi rubbed two sticks together to make fire, as his forebears had done even before they took our present human form. We have always known in our bones that we can create heat by doing work. Caloric theorists attempted to fill this one gap with arguments about how friction released surface caloric, or how caloric could be liberated from a material by kneading it.

Whether they dealt with Aristotelian essences, phlogiston, or caloric, philosophers and scientists looked at frictional heating and saw not a contradiction but rather a phenomenon that demanded explanation in terms of their particular subtle fluid. All the while, they had to have known in their gut what their heads were slow to admit—that we have always been able to create caloric, as long as we did not tire of doing so.

The stage was thus set for the last act in the drama of writing the new science of thermodynamics. It would be necessary to fully digest the fact that thermal energy and mechanical work can be traded back and forth. That would be the essence of the first law of thermodynamics—the law of conservation of energy.

While the very presence of steam engines was forcing people to face the fact that thermal energy could be turned into mechanical work, that fact had an obverse: the production of heat by friction. Using heat to create work required complex machinery, but turning work into thermal energy was extremely easy. Once we recognized the action of friction as energy *conversion* instead of energy *liberation,* it would finally be possible to explain the steam engine—formulating the new science of thermodynamics along the way.

Physicist Paul Epstein pointed out that the frictional liberation of heat had caused physiologists to take the interchangeability of heat and work for granted long before it was accepted by people in the physical sciences.[10] He introduces the fascinating eighteenth-century physiologist Albrecht Haller. Haller was born in Switzerland in 1708.[11] His genius was soon apparent and, as with so many smart people, his terrible need for approval surfaced early and lingered long.

Haller finished medical school at Leiden when he was nineteen. He traveled some, then returned to Bern to look for a professorship, not in medicine but in history and rhetoric. He did some lecturing on anatomy and medical practice, but he also walked the mountains writing poetry. The poetry may have been less than great, but Germany had yet to give us great poets. Haller's first published volume of poems in 1732 proved to be very popular. The collection went through many editions.[12]

The great transition in Haller's life came at the age of twenty-eight. He applied for and was given a post in the medical school at the University

XIX. Trauer-Ode,
beym Absterben
seiner geliebten Mariane,
Nov. 1736.

Soll ich von Deinem Tode singen?
O Mariane! welch ein Lied!
Wann Seufzer mit den Worten ringen,
Und ein Begriff den andern flieht.
Die Lust, die ich an Dir gefunden,
Vergrössert jetzund meine Noth;

From the ninth edition of Haller's poetry, part of a sad ode to his first wife, Mariane, who died the year he moved to Göttingen.

of Göttingen. During his seventeen years there he set the modern form of the subject of physiology. His monumental text on the subject was still a major medical source book in the mid-nineteenth century.[13]

Haller revolutionized our knowledge of blood flow and heart action, clarified the relation between respiration and blood flow, and explained nerve action in muscles. He gave us new insights into human reproduction and birth defects.

Then, at forty-five, he abruptly quit his academic job to accept the modest post of court bailiff in Bern. The greatest physiologist of his age went back home to count votes. That makes sense only in the light of the inner torment that seems to have permeated Haller's entire life. He had an instinct for destroying friendships and reacting violently to small slights. He appears to have been tormented by religious doubts and self-incrimination—all the time craving acceptance.

Here, at last, was a form of acceptance right in his hometown (even though the final selection of the person for that post was done by lot). Haller left a litter of broken friendships in Germany. However, he did good work in his new post and then subsequently took charge of a local salt works. His biographer Lester King quotes an earlier biographer (Henry Sigerist) who called his salt works post "a little kingdom of his own which he studied and which he ruled wisely."[14] Haller wrote more physiology, but he never went back to the university. He also wrote several novels, which, like his poetry, were well received.

Along with all the things Haller had gotten right in his physiology, he got one thing both right and wrong at the same time. In his explanation of how our blood is heated, he said:

> We may also ask, whether the heat of the blood does not also proceed from its motion? [Seeing] that we observe heat to arise from the motion of all kinds of fluids, even of air itself, in our experiments; but much more does this attrition produce heat in the inflammable animal juices, which are denser than water, and compressed with a considerable force by contractile and converging tubes.[15]

Haller expands upon that idea later in the book when he talks about the action of the lungs:

> It may be asked therefore, whether the blood does not acquire its heat principally in the lungs? . . . The lungs . . . will add to the office of the rest of the arteries, because in them the blood is alternately relaxed and compressed more than in any other part of the body.[16]

Both statements seem utterly reasonable to our present-day thinking, even though they are actually incorrect. Our blood temperature is, as it turns out, kept up by chemical combustion, not by mechanical action. Yet in these statements Haller perfectly represented the fact that mechanical friction does indeed produce thermal energy. He did so matter-of-factly, without making any cosmic inferences about the nature of things. He did so as one person reporting to another an idea that both took for granted.

The notion that mechanical action heats blood not only survived but gained momentum. We see this in a prominent Haller follower, the charismatic Colonial doctor (and Declaration of Independence signer) Benjamin Rush. Rush was highly influential as a teacher of medicine. He focused on the role of arteries in fever.[17] He believed that fevers rose when the pulse and arteries became overactive due to irritation and began to mechanically overwork the blood. He therefore attacked the resulting temperature rise by heavy bleeding of patients. Rush is often credited as being a father of psychiatry since he wrote a good deal on the subject of mental illness.[18] (He found that fevers were associated with some manic behavior, and therefore used bleeding on mental illnesses as well as other diseases.)

Gradually, however, it became clear that chemical reactions occurred in the blood and that those reactions contributed to the maintenance of our body heat. That recognition began with Priestley and Lavoisier. Epstein traces the subsequent history of blood heat through to its resolution in 1843. At first, various estimates of the extent of the chemical heating were all low. They failed to account for all of the blood heat.

During the early nineteenth century, the caloric theory continued to overshadow the mechanical theory of heat. Yet physiologists kept attributing a fraction of blood heat to friction. By 1843, French chemist Pierre Dulong had access to better information and could finally show that chemical heating accounted for *all* of blood heat. The caloric theory was by then embattled but still ascendant. The fact that frictional heating now vanished even from physiology bought it some additional time.

Everyone who has ever studied the history of heat has struggled with the obviousness of mechanical friction. Whether it was noted explicitly, as by physiologists, or explained away by caloric theorists, people had constantly been aware of it. The seventeenth-century philosopher John Locke, for example, had commented on how the portion of an axle where the wheel rubbed against it grew hot under the resulting friction.

As long as Haller's explanation of how our blood stayed warm lingered, it never did assume any role as a competitor for caloric. Nor did the many people who had mentioned the kinetic origins of heating rise as opponents to the general notion of a subtle fluid. So we must ask when the mechanical theory *did* take its place as a competing explanation of heat. The answer appears to lie in the person of the American expatriate Benjamin Thompson.[19]

Thompson was born in Massachusetts in 1753 and raised in Woburn. He wrestled out a patchy education in Boston and at eighteen, just before the American Revolution, went off to Rumford, Massachusetts, as its new schoolmaster. He soon married a well-to-do thirty-one-year-old Rumford widow. After a while it became clear to people in the region that he was also a Tory, spying for the British. At that point, he deserted his wife and a new daughter and fled to England. There he worked with the admiralty, did ballistics studies (for which he was made a member of the Royal Academy), returned to America as a British officer in the American Revolution, and finally went to Europe, where he, like Becher before him, was engaged in Munich as an advisor to the elector of Bavaria.

His life took on new coloration in Munich. He instituted social reforms that were years ahead of their time. He devised public works, military reforms, and poorhouses. He equipped them with radical kitchen, heating, and lighting systems. He created the spacious and beautiful People's Park within the city of Munich. (It remains a major attraction there today.) And yet, politically, he was completely royalist. He had no use for populism in any form. For Thompson, social reform was merely a matter of making the country run more smoothly.

In 1792, he was made a count of the Holy Roman Empire, and he took the name of the town he had once fled—the name of Rumford. (That same year, his estranged wife, Sarah Rolfe Thompson, died.) As a result, the name we remember for foundational work on the mechanical theory of heat is Count Rumford, not Benjamin Thompson, despite the fact that he had already spent years working on thermal questions.

Rumford had even dispatched a spy, a young mechanic named Georg Reichenbach, to England the year before he was formally made a count. Reichenbach got inside the Boulton-Watt factory in Birmingham, where he combined memory with a fine sketching ability. He brought plans for a Watt engine back to Thompson, who for a while thought about manufacturing steam engines.

But that task lay beyond even his vast energy and intelligence. To make matters worse, Watt got wind of the deception, and Rumford was left only with Watt's lifelong wrath. However, in the wake of his steam engine adventure, he moved on to the study that eventually would do more than any other to clarify how a steam engine really worked.

Rumford now moved from studying stoves and domestic heating systems to studying the nature of heat itself. As Count Rumford, he made a series of very insightful experiments with field artillery. He found that when he bored a cannon barrel under water with a blunt bit, he could heat the water to boiling and then keep it there as long as he wished. Instead of cutting metal, he created continuous friction and continuous heat production. That result made it perfectly clear that this was no liberation of caloric, for there was no limit.

Rumford also did another, quite different kind of cannon experiment. When he fired cannons with and without cannonballs, the ones fired without cannonballs became far hotter. If the cannon were loaded with a ball, the explosion did work accelerating it. The ball carried kinetic energy away with it, and less energy remained in the form of heat.

Finally, as a consequence of his experiments, Rumford was able to state quite plainly:

> Anything which an insulated body, or system of bodies, can continue to furnish *without limitation* cannot possibly be a *material substance;* and it appears to me to be extremely difficult, if not quite impossible, to form any distinct idea of any thing, capable of being excited and communicated in the manner the Heat was excited and communicated in these experiments, except it be MOTION.[20]

When these results were published in Great Britain, they finally gave rise to an anti-caloric faction. Caloric theorists counterattacked by asking Rumford how he could speak of furnishing "energy without limitation" when he had run his experiments over finite periods. The caloric theory would prevail for another half century, but at least the assault upon it was now out in the open.

Soon after he did his cannon experiments, a now wealthy Rumford returned to a life lived between London and Paris. He was a founder of the Royal Institution of Great Britain. He went back to America to re-unite with his daughter, Sarah (now the Countess Sarah Rumford). In a remarkable bit of historic irony, Rumford also began a friendship with Marie Lavoisier (widowed when Antoine Lavoisier was guillotined during the French Revolution). Their intense affair ran for four years, and they finally married in 1805. Before the marriage Rumford crowed:

> I think I shall live to drive *caloric* off the stage as the late M. Lavoisier (the author of caloric) drove away Phlogiston. What a singular destiny for the wife of two *Philosophers!!*[21]

Still, Rumford took pains to describe Lavoisier's work with utmost respect, gently leading readers to the impression that the great scientist had never really believed *calorique* to be a material fluid. Whatever Antoine Lavoisier had believed, Marie Lavoisier (who was very scientifically savvy) took Rumford's side in the ensuing debates over caloric. Rumford biographer Sanford Brown even speculates that she might have egged Rumford on.

But then they married, and that pretty much spelled the end of Rumford's story. The marriage itself was nasty, brutish, and short. When they parted three years later, he was calling her a "female dragon." It was a clear conflict of lifestyles. He, for example, set out to redesign and modernize her house, while she wanted to continue as a major Parisian social figure in the style she had known during the ancien régime.

After that, Rumford continued his interest in laboratory experiments, but otherwise he increasingly withdrew his flamboyant social presence until his death in 1814. The caloric question would remain unresolved for several more decades.

By the early nineteenth century, it might seem that we were close to putting caloric to rest once and for all, along with all the other versions of the essence of fire. However, a subtle advance in our understanding of heat had the surprising effect of postponing that recognition for a generation.

This is where a French father and son enter our thermodynamical stage. They are Lazare Carnot and his son Sadi Carnot. Lazare Carnot was one of the truly remarkable figures in the eighteenth century, although his son and grandson were better known than he.[22] Son Sadi Carnot gave us the first incarnation of the second law of thermodynamics. Lazare's grandson and Sadi's nephew (who was also named Sadi Carnot) became the president of France from 1887 until he was assassinated in 1894.

Lazare was born in 1753 in eastern France, the same year Benjamin Thompson was born in America. Carnot finished an education in mathematics and military engineering and went into several years of military service. During that time, he competed for prizes in mathematics. He also had political dealings with the infamous Robespierre.

Then he became involved in an affair with an aristocrat's daughter. Finally her father, unbeknownst to Carnot, arranged her marriage to another aristocrat. Carnot furiously went to the fiancé and revealed the affair which broke up the marriage plans, and the father had Carnot thrown in jail. That was in the spring of 1789. The first events of the French Revolution were just taking place, and those events led to Carnot being retrieved from prison after only two months.

His life had been rather static up to this point. Now it began moving. He skirted the more radical actions of the Revolution and in 1796 was made a member of France's five-man ruling group, called the Directory. It was they who reorganized the government and ran it until Napoleon took power. And it was Carnot who started Napoleon on his rapid ascent to power by appointing him head of the Army of Italy.

Then a coup in the Directory forced Carnot into a brief exile in Switzerland. When Napoleon took over, Carnot returned and did many forms of high-level government service until Napoleon's regime collapsed. After that, the returning monarchy remembered that Carnot had once voted to behead Louis XVI, and he spent the rest of his life exiled to Germany.

Lazare Carnot was multitalented. He was an excellent violinist, and although he had a mathematical center of gravity, he was strongly interested in technology. He befriended people such as the Montgolfier brothers, as well as Robert Fulton, who showed up in France trying to sell submarine designs. At his core, Carnot thought like a technocrat. He once remarked:

> If real mathematicians were to take up economics and apply experimental methods, a new science would be created—a science which would only need to be animated by the love of humanity in order to transform government.[23]

The eighteenth-century French aristocracy had embraced Newton over his rival Leibniz, but Carnot came down on the other side. He not only used Leibniz's calculus but also adopted, and elaborated upon, Leibniz's idea that mechanical energy must be conserved. He turned his attention to the production of power.

He pointed out that in an imaginary perfect waterwheel, none of the water's energy would go to waste. None would be dissipated, and all the motion would be completely reversible. Water would be stationary before it entered and it would be stationary at the exit. Then he added an extremely important insight: if the perfect waterwheel were run backward, it would become the perfect pump.

This is where Lazare's brilliant son Sadi claimed his inheritance.[24] At the age of only twenty-eight, Sadi Carnot wrote his sole published work, a monograph titled *Réflexions sur la puissance motrice du feu (Reflections on the Motive Power of Heat)*.[25] In it, he suggested that we conceive a perfectly reversible steam engine. If we could build such a machine, we could run it in reverse and pump heat from a low-temperature condenser to a high-temperature boiler.

Refrigerators would not appear until thirty-six years later, but when they were built, they would be exactly what Sadi Carnot described. They would be reversed heat engines. And Sadi had just begun. He "operated" his perfect engine in a thought experiment. In his mental engine, he used an ideal gas instead of steam, and he imposed the constraint that no engine could possibly act as a perpetual motion machine.[26] He found that the extent of work one kilogram of air would produce in such an engine depends only upon the temperature at which the air is heated and that at which it is cooled.

This was the basis for a conclusion that we now call Carnot's theorem. It said that the motive force of a perfectly reversible engine depends solely upon the high temperature and the low temperature (the boiler and condenser temperatures in the case of a steam engine). That would be true whether the engine used steam, air, or any other fluid. It would be, in that sense, very similar to the output of his father's perfect waterwheel—a waterwheel that depends solely upon how far water falls through it. That primal dependence upon temperature difference proved to be the first step in stating what has become the second law of thermodynamics.

A curious fact about all this is that both father and son accepted the caloric theory. Indeed, I can find no evidence that Lazare Carnot and his contemporary Count Rumford ever communicated. (Perhaps that is because Rumford had advocated doing away with the caloric theory.)

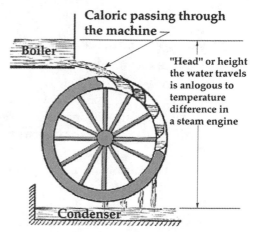

Carnot's steam engine/waterwheel analogy

Sadi Carnot still assumed that caloric was conserved as it passed through an engine, just as water is conserved as it passes through a waterwheel. What really happens is that heat from the boiler is partially converted to useful work, and only a fraction of it passes into the condenser. Since Carnot had not yet adopted the mechanical theory of heat, the validity of what he had said about steam engine performance appeared to bolster the caloric theory. This was a truly strange turn of affairs.

Caloric was thus still ascendant on February 22, 1840, when twenty-five-year-old Robert Julius Mayer shipped to the East Indies as the surgeon aboard the Dutch vessel *Java*.[27] Mayer had studied medicine at the University of Tübingen and taken longer than most to finish his M.D. degree. He had been something of a playboy; the only subject that had managed to compete with his interest in cards and billiards was anatomy. After he finished school, he spent another year gaining a Dutch license to practice medicine. Now he was the ship's doctor and bound for Batavia, in Java.

Here we find an echo of Haller and Rush, for when Mayer reached Java, a passing observation changed his life. Part of his work involved letting blood from feverish sailors. To do that, one always lanced a vein and avoided the arteries. Since venous blood has spent its oxygen, it normally runs darker than arterial blood.

The first time Mayer opened a vein in Djakarta, the blood ran far too red—so red that he thought he had hit an artery. Then other physicians told him that was normal in the tropics. Mayer realized that people in such hot climates burn less of the food they eat, since they need to generate less heat.

Mayer thought about that red blood on the long voyage back to Holland. It was obvious that, by some means, food fuels our power output just as, by some means, coal fuels a steam engine's output. Mayer now realized that food also fuels our heat supply. He had been bored on the trip out. But on the return voyage he appears to have grown remote and gone into a brown study, not even taking full shore leaves when he had the chance. He put the two pieces together. If food provided both heat and work, work and heat must be linked.

Back in Germany in 1841 he submitted a paper on his ideas about energy equivalence, but he knew too little mathematics or physics. The paper, laden with misconceptions, was simply ignored. So he studied more basic science and, a year later, he wrote a better paper. This one was published, but it too was ignored.

Mayer meanwhile picked up what appeared to be the threads of normalcy. He married, and he appeared to be settling into the everyday life of an early-nineteenth-century physician—first in private practice, then in the role of chief town surgeon. But it was not to be. The creative daemon had its claws in Mayer, and it was not about to release him.

He kept honing his ideas, and in 1845 he self-published his next paper—a long monograph that proved to be his major work. In it, he used existing ideal gas data in which, he realized, were already embedded measurements of the heat/work equivalence. Mayer used a completely correct calculation to get a work/heat equivalent that was within

Robert Julius Mayer (1814–1878).

86 percent of the correct value. (His result was expressed in units we would not use today. He got 365 kg-m/kcal for the work/heat equivalency.) The modest error stemmed from the fact that the data he used were still imperfect.

Although Mayer had spun a completely correct theory, others would not believe it without its being backed up by additional measurements. Mayer's frustrations mounted: during the time he came under attack for his scientific studies, three of his five children died in infancy. Insight had laid its hand upon Mayer in Java, but professional scientists—people untouched by the daemon, you might say—had to see Mayer's work put into familiar terms. By 1850 Mayer was angry and frustrated. He attempted suicide and was for years in and out of asylums.

Meanwhile, in Manchester, England, another young man, four years Mayer's junior, took up science as a hobby. He was James Prescott Joule, son of a brewery owner.[28] Joule's father hired no less a person than John Dalton as his son's tutor. (Dalton had by then originated the concept of atomic number and the first rules of chemical stoichiometry.) Joule proved to be a very apt pupil. As early as 1840 he had begun studying the generation of heat that occurs when electricity flows through a resistor.

He was only twenty-two when he showed that the heat released in an electrical resistor depends upon the product of the resistance (ohms) and the square of the current (amperes) flowing through it—and upon nothing else. Today we honor Joule by naming the basic unit of energy after him. In the International System of units (upon which all of our units in the United States are based) one joule is one newton-meter of work.

So Joule recognized the direct connection between work and heat. He next began a series of experiments in which he drove a paddle wheel in a closed insulated container of water. With great accuracy, he measured the work that he did turning the paddle wheel, and the resulting rise in the temperature of the water. He continued polishing the precision of these and other similar measurements until, in the year 1847, he was able to present an equivalence between foot-pounds of work and British thermal units. His data lay within 99 percent of the correct value, which was later determined to be 778 ft-lb/Btu.

Like Mayer, Joule lacked the formal background that would have greatly extended his capacity for capitalizing upon his own accomplishments. He first presented his thermal equivalence measurements in May 1847 in the reading room of St. Ann's Church in Manchester, and the text of his talk was published in the *Manchester Courier.*

Joule's apparatus for measuring the temperature rise of water churned by a paddle wheel (London Science Museum).

Of course Mayer and Joule—functioning as theorist and experimentalist—collided. Joule became one of Mayer's early critics. When Joule's work appeared in the *Comptes Rendus* of the Paris Academy of Sciences later in 1847, it contained a complaint that one of Mayer's assumptions could only be justified after the fact by Joule's own data.

Other issues swirled between them. Each had different ideas about how or whether mechanical motion was sustained once it had been transmuted into heat. And Carnot's work, which had been given little initial attention, now began to bother people. It was clearly valid even though, based upon its use of caloric, it appeared to contradict Mayer and Joule.

The subject of thermodynamics, as we understand it today, would come about in resolving this crisis, and the person who might best be credited with its formulation was Rudolf Julius Emmanuel Clausius.[29] If anyone is to hold the title of canonical inventor of thermodynamics, I suggest this would be the person to do so.

The Prussian Clausius was born in 1822. In 1847 he received a doctorate in mathematics and physics from Halle University in Wittenberg. He published his first study of heat in 1850 and by 1865 had laid the formal foundations of the subject of thermodynamics much as it is still taught.

The relationship of Clausius to thermodynamics is similar to, say, the relationship of the Wrights to the airplane. The primary pieces of the puzzle were finally all in place. Clausius assembled the conceptual elements of thermodynamics much as the Wright brothers finally assembled the conceptual pieces of a successful airplane.

Clausius gave thermodynamics its first axiomatic structure. He synthesized a new science from seemingly contradictory parts. In particular, he showed how Carnot's theorem and the mechanical theory of heat complemented each other. In accordance with the mechanical theory, less heat left a steam power cycle than entered it—the difference being heat converted into useful work. That much contradicted Carnot, but it did not negate Carnot's major achievement.

Clausius saw that something *was* being conserved in Carnot's perfectly reversible engine; it was just something other than heat. Clausius identified that quantity, and he gave it the name *entropy*. He found that if he defined entropy as the heat flow from a body divided by its absolute temperature, then the entropy changes in a perfectly reversible engine would indeed balance out. As heat flowed from the boiler to the steam, the boiler's entropy was reduced. As heat flowed into the condenser coolant, the coolant's entropy increased by the same amount.

No heat flowed as steam expanded in the cylinder or, as condensed water, it was compressed back to the boiler pressure. Therefore the entropy changed only when heat flowed to and from the condenser and the boiler and the net entropy was zero.

If the engine was perfectly reversible, it and the surroundings with which it interacted remained unchanged after each cycle of the engine. Under his definition of entropy Clausius was able to show that everything Carnot had claimed was true, except that heat was conserved in his engine.

Once Carnot's work had been relieved of that single limitation, Clausius could reach another important result: the efficiency of a perfectly reversible heat engine depends upon nothing other than the temperatures of the boiler and the temperature of the condenser. The efficacy of the engine thus really does depend upon nothing beyond temperature, just as Carnot had said it must—even though his use of caloric denied him the specific use of the word *efficiency*.

But now we drift into the language of formal thermodynamics instruction, and that is not where we want to go. The point is that Clausius was able to show that irreversible engines—imperfect real engines—increase

Clausius's demonstration of Carnot's theorem can easily be made with a little arithmetic. First, he defined the *efficiency* of Carnot's engine— something that Carnot (without accepting the equivalence of mechanical and thermal energy) could not do. The efficiency is the ratio of the work done to the heat received in the boiler. For any engine, including Carnot's perfect engine, the efficiency is:

$$\text{Efficiency} = \text{Work}/Q_b$$

where Q_b is the heat from the boiler. If we recognize that the work is equal to the heat received minus the heat rejected, and if we identify the heat passing into the condenser as Q_c, then

$$\text{Efficiency} = (Q_b - Q_c)/Q_b$$

Now let us call the absolute temperatures in the boiler and condenser T_b and T_c, and the entropy changes (denoted with the symbol ΔS) in each are ΔS_b and ΔS_c. Bearing in mind that ΔS_b must be equal to ΔS_c, we get:

$$\text{Efficiency} = (T_b\,\Delta S_b - T_c\,\Delta S_c)/T_b\,\Delta S_b$$
$$\text{or}$$
$$\text{Efficiency} = (T_b - T_c)/T_b$$

the net entropy in their world rather than leaving it unchanged. Eventually we would be able to show that his entropy was a reflection of atomic disorder. Atoms, however, were still a small presence on Clausius's horizon. For now, Clausius had finally assembled all the pieces of a non-atomist description of heat phenomena. Clausius had codified thermodynamics.

Let us return for a last look at Mayer and Joule, since they exemplify so much that is mischievous about priority. When Clausius finally resolved the subject, Mayer was overlooked. But, unlike many brushed-aside inventors, Mayer's name was rehabilitated during his own lifetime by the great Irish/British scientist John Tyndall. In 1863, Tyndall wrote an important text, *Heat: A Mode of Motion.*[30] The book began and ended with Mayer's contribution. Even more important, Tyndall recognized and articulated the subtle way in which Joule and Mayer complemented each other. Tyndall wrote, "In the firmament of science Mayer and Joule constitute a double star, the light of each being in a certain sense complementary to that of the other."[31]

Mayer wrote to thank Tyndall. He said, "The hopes which in silence I ventured to cherish were more than fulfilled by the recognition which you there accorded me." In that same letter he says of Joule, "I have never regarded him as an antagonist, but . . . an esteemed and renowned fellow-labourer in the same domain of thought."[32]

The Laws of Thermodynamics

The classical thermodynamics that evolved upon the work of Clausius was a structure of axioms and the deductions that flowed from those axioms. Many such axiomatic formulations exist, and this particular set of axioms I chose for its close kinship with Clausius's original ideas.

The first law	Energy can neither be created nor destroyed.
	Or: The energy of the universe (or of any isolated system) is constant.
The second law	It is impossible to build any heat engine operating between two heat reservoirs (such as a boiler or a condenser) with an efficiency greater than a Carnot engine.
	Or: It is impossible to create any heat engine that has no effect upon the universe other than cooling a single heat reservoir and lifting a weight.
	Or: The entropy of the universe (or of any isolated system) can, as the result of any process occurring within it, only stay constant or increase.
	(These three statements can be proven to be completely consistent with one another.)
The third law	The entropy of any substance vanishes at a temperature of absolute zero. (This law dates to the early twentieth century, and its exact form is still the subject of some debate.)
The zeroth law	Any two bodies in thermal equilibrium with a third body are in thermal equilibrium with one another. (This fact, recognized by Joseph Black, is necessary to complete the axiomatic structure.)
The state principle	Two independent thermal properties are sufficient to completely specify the state of a simple compressible substance. (This is very necessary to the logical structure and was long taken for granted in thermodynamics. I have stated it in a rather simple and restrictive form here.)

Tyndall thus created peace in troubled waters by cutting through to the essence of the conflict. In the preface of his book, he demonstrated a keen insight into the texture of the long process that had finally produced this fine intellectual edifice. Tyndall offers a passage that not only encapsulates the conflict between Mayer and Joule but also diagnoses the conflict that has torn at scientific discourse right down to the present day:

I have therefore tried to show the tendency displayed throughout history, by the most profound investigators, to pass from the world of the senses to a world where vision becomes spiritual, where principles are elaborated, and from which the explorer emerges with conceptions and conclusions, to be approved or rejected according as they coincide with sensible things.[33]

In other words, Tyndall acknowledged the twofold path to understanding represented by the visionary Mayer and by the fastidious experimentalist Joule.

In another passage, Tyndall offers an extraordinary reading of the first law of thermodynamics, which had been, after all, the great hurdle that had to be overcome before we could complete the intellectual task the steam engine had presented to us:

> The law of conservation rigidly excludes both creation and annihilation. Waves may change to ripples, and ripples to waves—magnitude may be substituted for number, and number for magnitude—asteroids may aggregate into suns, suns may invest their energy in florae and faunae, and florae and faunae may melt in air—the flux of power is eternally the same. It rolls in music through the ages, while the manifestations of physical life, as well as the display of physical phenomena, are but modulations of its rhythm.[34]

It had taken the steam engine about 150 years to drive us to the completion of the invention of thermodynamics, and Tyndall now took a moment to take pure pleasure in what had been achieved.

Let us turn next to another invention, if you will allow me to call it that—an invention that flowed quite naturally from the invention of the steam engine. It would be natural to go next to the invention of the automobile, the locomotive, or any other powered conveyance. But to understand any invention, we must understand what drives it. And I suspect that what propelled us to make the nineteenth century the age of powered conveyances was something more than a simple need to get from here to there.

We turn next to the "invention" of speed.

> Arrest my heart, my brain will beat as true;
> and if you set this brain of mine afire,
> then on my blood I yet will carry you.[35]
>
> —Rainer Maria Rilke

7

Inventing Speed

A picture in an 1866 French book on heat offers a curious glimpse of a mentality that followed in the wake of Watt.[1] The author, Achille Cazin, said he had no interest in high-flown theories of heat—he would deal only with heat phenomena. It is a limited objective that he carries out very well. The first image in his richly illustrated book shows what he calls an *éolipyle*, the French version of the word *aeolipile*. It is a jet-propelled tricycle that fairly screams speed.

Fig. 1. — Éolipyle à recul.

Cazin titles his picture "Recoiling Aeolipile," and we need to look at each of those two words. The word *aeolipile* had generally been attached to Hero's first-century turbine (Chapter 4). In Cazin's nineteenth-century mind, Hero's machine becomes a propelled vehicle; perhaps we should not be surprised by that. By this time it had become very clear that the motive force of steam was wonderfully suited to driving vehicles.

A vehicle very similar to this one had been seriously suggested 145 years earlier by the Dutch Newtonian philosopher Willem Jacob 'sGravesande, a student of the concept of momentum.[2] But if Cazin had read 'sGravesande's work, he appears to have missed the point. A small detail in the image, one that is easy to overlook at first glance, reminds us that he is still fuzzy on how the machine should work: a cork, riding in the jet, flies off the far side of the picture.

Cazin says that the trick is to raise the steam pressure in a corked spherical tank until the cork suddenly explodes out of the nozzle like a bullet. He stresses the gun-and-bullet analogy. And here we come to the second word, *recoiling*. This device certainly does represent recoil propulsion, but the cork has little to do with the recoil. You and I understand that the jet, not the cork, drives such a machine.

The steam turbine concept—blades driven by steam jets—was already on many inventors' minds, though it would be another quarter century before functioning turbines were on the market. Jet engines, working on the principle of recoil (or reaction, as we now call it), would appear later still. Jet engines and turbines would eventually be the basis for very high speeds. By the late nineteenth century, however, steam engines were already driving some very fast vehicles.

The nineteenth century was the epoch in which the newly realized capabilities of gases would finally answer our age-old yearning for speed. For all of recorded history we had wanted to capture the sensations we had known only for brief instants. We had always dreamt of swinging through the air—of running ahead of the wind. We had always wanted to shake off the limitations of our bodies—to leave the eagle and the gazelle behind us.

Watt's fully matured steam engine and the Mongolfier brothers' first human-carrying balloon had appeared within months of one another. Gases were poised to become the fifth essence, the empyrean essence, that would make all things possible.

Tracing antecedents of the locomotive or the airplane can cause us to overlook the real need being fulfilled. Individual inventors would go their various ways, but a common wish (our old friend the Zeitgeist) was driving these technologies. Behind all the new powered vehicles that began appearing was the realization that machines could now give us the winged feet we had always longed to have.

One harbinger of nineteenth-century speed actually had a great deal in common with Cazin's aeolipile. When, during the War of 1812, Francis Scott Key wrote about "the rockets' red glare, the bombs bursting in air," he was actually reacting to the drama of an old Asian technology. For Great Britain had brought the ancient technology of war rockets to maturity. The thirty-two-pound Congreve rockets that fell upon Fort McHenry in 1814 had become a serious weapon of war, with a range of about three miles.[3]

For a time—during the early nineteenth century—military rockets were widely used throughout Europe. But they had originated in China almost a millennium before, and their use had spread to other countries as well. War rockets appeared sporadically in Europe during the Renaissance but failed to take root.

They did, however, find regular tactical use in India, whose landscape of small hills and stony riverbeds made it hard to move artillery. India also had the best Asian supplies of the saltpeter needed to make powder. The sixteenth-century Mogul emperor Akbar used military rockets, and by the eighteenth century rocket troops were a part of most Indian armies, which used six- to twelve-pound rockets with a range of a kilometer or so. Those Indian rockets caused considerable disruption to invading English cavalry, but they might have remained in India were it not for a bright young man named William Congreve.

Congreve, born in 1772, the son of a British general who served in the American Revolution, became the comptroller of the Royal Laboratory at Woolwich. Young Congreve grew fascinated by the Indian rockets that were showing up in the Royal Artillery Museum. The wonderfully inventive Congreve had already published a folio of his many inventions in 1800. Then, in 1804, he realized that rockets exerted no reactive force on a rocket launcher—they had none of the "kick" of a cannon. He wrote, "It first occurred to me, that, as the projectile force of the rocket is exerted without any re-action upon the point from which it is discharged, it might be successfully applied, both afloat and ashore, as a military engine."[4]

Congreve developed a series of rockets a good deal more destructive than those in India. Two years later, British warships began using them in battle. After testing them against French shore installations, they fired a barrage of 25,000 Congreve rockets at Copenhagen—a hapless bystander in the Napoleonic Wars. The rockets raised havoc by starting fires throughout much of the city. From then on, British rockets played a continuing role in their war with the French.

During the War of 1812, Great Britain turned her Congreve rockets on America—all the way from Bangor, Maine, to New Orleans. During the campaign in which they sacked Washington, D.C., the British also tried to take Baltimore. But Fort McHenry, which guarded the city, managed to

Part of a Congreve rocket fired during the Battle of Stonington, August 9–12, 1814, by the British ship *Terror* (Stonington Lighthouse Museum, Stonington, CT).

endure the rockets. It turns out that the "bombs bursting in air" proved more effective than the rockets, whose red glare made fine theater but did less damage to stone forts than to wooden cities.

War rockets lasted another forty years. They were finally replaced with heavy rifled artillery on ironclad ships and did not reappear until such ordnance as American bazookas and German V-2 rockets came into use during World War II. Yet if they came and went with the Industrial Revolution, they did manage to whisper their message of speed before they left. They hinted that forces exerted by gases would one day move us with blinding speed. But they left us with their huge plodding steam engine cousins. Those engines would have to be transformed before they could provide us with the speed we began to crave.

At the beginning of the nineteenth century, humans had moved on land with the speed of a horse, or on sea with the speed of a sailboat. They had briefly moved pretty rapidly on ice. For a short time, a fast human runner might challenge any of these, but that was all we could manage. Our most poignant experiences of speed had all occurred in short bursts—diving, swinging on a rope, and (primarily) riding fast horses.

Although a horse is not really much faster than a person, to ride upon one in full gallop—to be conjoined with an animal eight times your weight and moving at speeds that can literally be called breakneck—is pretty heady stuff. Bayard Taylor was only one of many nineteenth-century writers and poets who captured the sensate force of the horse in motion. In 1853, he began his "Bedouin Song" with these lines:

> From the desert I came to thee
> On a stallion shod with fire,
> And the winds are left behind
> In the speed of my desire.[5]

We would soon enough learn to leave the winds far behind, for the horse had merely teased us—taunted us with as much as forty miles an hour, but only for a minute or two at a time.

That desire now began permeating our language. In his 1841 essay "Prudence," Emerson recognized that our fixation on speed went far beyond simple speed of movement. Rapidity in all its forms had become a Yankee trait—perhaps even a virtue:

> Our Yankee trade . . . takes bank-notes,—good, bad, clean, ragged,—and saves itself by the speed with which it passes them off. Iron cannot rust, nor beer sour, nor timber rot, nor calicoes go out of fashion, nor money stocks depreciate, in the few swift moments in which the Yankee suffers any one of them to remain in his possession. In skating over thin ice, our safety is in our speed.[6]

We Americans were becoming like spit on a hot stove, always in motion. Speed on fast horses had been enticing because it was dangerous. Now, in a peculiar inversion of that idea, speed was presented as protection from the dangers that were inevitably all around us in our raw new land.

No doubt rapidity did offer means for eluding danger. But the danger that attended speed had, of itself, become a powerful lure—tempting us to skate over all kinds of thin ice that we would otherwise had avoided. Three years after Emerson's essay, the American diarist Philip Hone sounded a note of regret at all this escalating speed. In his entry for November 18, 1844, he wrote, "By and by we shall have balloons pass over to London between sun and sun. Oh for the good old days of heavy post-coaches and speed at the rate of six miles an hour!"[7]

Also in 1844, as the potential of the new steam cars and railways was growing increasingly evident, artist J. M. William Turner captured this impulse toward speed with a fine visual metaphor. In what might be his most famous painting, *Rain, Steam, and Speed—The Great Western Railway,* Turner offers an abstract vision of a locomotive hurtling over a bridge in a storm. We squint through the slanting rain to see it. The engine appears to be a machine before its time—an anticipation of things to come.

The painting implies speed yet to be realized. The train in the center is dark and rain-shrouded. On either side is a golden landscape, pastoral and rustic. Art historians have spilled a sea of ink upon that extraordinary picture. It is impressionist before the impressionists. It is a joyous celebration of our new technological power at the same time it threatens terrible disruption to its world as it passes through.

Other great artists joined Turner in displaying these speeding engines as agents of upheaval. In 1855, American artist George Inness did an epic canvas of a primitive locomotive spewing smoke across the pastoral Lackawanna Valley. Tree stumps and distant smokestacks signal a fast-changing world. By 1880, Monet, Pissarro, Manet, and Degas had all shown us their impressions of a way of life transformed by locomotives. In these later works, however, it is the train station that emerges as the new center of modern life, and the menace ebbs.

When Jules Verne wrote *Around the World in Eighty Days* in 1872, he told how his protagonists Phileas Fogg and Passepartout make their journey at the surprisingly low average speed of around 10 miles per hour. As the urge toward speed was being realized, somehow it was also being tempered by realism. But all that was later. An 1831 work done by the English painter Thomas Talbot Bury shows a truly embryonic train slicing across the patchwork fields of England on a long, straight, willful line of track. Clouds gather over it; Turner's rain is about to fall.

Rain, Steam, and Speed—The Great Western Railway, by J. M. W. Turner. This black-and-white reproduction of Turner's vision of speed was published in the same year the Wright brothers flew, 1903.

Of all these paintings, *Rain, Steam, and Speed* is the most disturbing. Turner shows us his indistinct and mystic train crossing a very real bridge. He placed it upon Isambard Kingdom Brunel's Maidenhead Bridge, and we can be fairly certain that's a sly conscious reference to the irrevocable change that the locomotive brought with it. The coming of speed clearly frightened some while it elated others.

What we invented in the nineteenth century was more than the sum of the parts: not trains, not steamships, not airships, but a quality shown by all. It was speed—pure, hedonistic, and inexorable speed. Speeds kept increasing until, in 1907, a *Stanley Steamer* was clocked at an unofficial but probably valid—and certainly astonishing—speed of 150 miles an hour.

We read so much about the invention of locomotives, automobiles, and airplanes. We naturally focus upon this technology or that, rather than on the human craving behind all the machinery. The nineteenth century was the one in which, after thousands of years, we finally "invented" speed. Let us therefore look at some milestones in that process.

Watt's engines were bringing his steam engine power-to-weight ratios to a point at which vehicles might more plausibly accommodate them. So although speed did not catch *his* fancy, it was certainly tempting others.

Since the largest existing vehicles were ships, they would become the first vehicles to regularly run on steam.

But experimental steam cars actually predate the first steamboats. Between 1769 and 1771, a Belgian named Nicolas Cugnot developed a three-wheeled, steam-powered caisson puller for the French army.[8] It could tow a three-ton cannon at about 2.5 miles per hour, but only for ten minutes at a time. The French withdrew support for it after the driver lost control and it crunched into a wall.

We read a similar failure to act upon steam's potential within the Boulton-Watt factory. It centers upon a very important company engineer, William Murdock (often spelled Murdoch). The statue commemorating the factory in Birmingham, England, gives us an idea of his importance. Along with Boulton and Watt stands Murdock. The three are engaged in a lively debate over a set of drawings.

In the early 1780s, Murdock worked on a series of small steam cars and seems to have been on track to produce successful models. To keep the size of their engines small he had to use fairly high-pressure steam.[9] Watt, however, distrusted the ability of the boiler manufacturers to contain such pressures; he saw no future for steam-powered vehicles. In 1784, he and Boulton patented such a vehicle—not to sell it, but to prevent others from trying to doing so. (Once again, we note the dubious role that patents can play in the inventive process.)

Steamboats were another matter. Even without Watt's engines, people had begun building them. The idea of putting a Newcomen engine in a boat had been alive from the beginning. As early as 1736, one Jonathan Hulls of Gloucester patented a steam tugboat. Whether or not he tried to build it, we are not sure. We do know that he was literally laughed out of town for his efforts.[10]

Hulls's steam tug as he represented it in his advertising notice.

Like the earliest known steam car, the earliest known steamboat was also built and run in France. And, like Cugnot's car, it was also a dead end. It began when two artillery officers, the comte d'Auxiron and Charles Monnin de Follenai, left the army in 1770 to work full time on a steamboat. By 1772 d'Auxiron had talked the government into promising the first successful builder a fifteen-year exclusive license to run the boat commercially.

Working in Paris, on an island in the Seine River, they installed a large Newcomen steam engine in a boat. Its low power-to-weight ratio was their undoing: the engine was so heavy that it sank the boat. After three years of lawsuits with stockholders and wrangling with coworkers, d'Auxiron died of apoplexy. But before he did, one of those coworkers, the marquis de Jouffroy d'Abbans, had picked up his work and taken it in a new direction.

At same time d'Auxiron had been getting started, the young aristocrat Jouffroy was thrown into a military prison on the isle of Sainte Marguerite for fighting a duel. During his long stay in jail, he had contemplated the traffic of ships and boats passing below, and he began drafting his own plans for building a steam-powered boat. When he was released from prison in 1775, he went off to find d'Auxiron and his supporters.

At first the group's hopes ran high for a new steamboat, but they disagreed on how much power a steamboat needed. High hopes turned to acrimony. A split took place, and Jouffroy led a new effort, far from Paris, in the relatively remote city of Lyon. The Paris people went ahead and bought a Watt engine, but then they rendered it nearly useless by "improving" it. Jouffroy, on the other hand, went ahead and built a Newcomen-type atmospheric engine of his own design.

By 1783, Jouffroy was able to make a public run in his 140-foot boat on the Saone River. Crowds along the shore cheered it for fifteen minutes, unaware that it was breaking up under the engine's pounding. Before anyone spotted trouble, Jouffroy managed to ease it to shore. He bowed to a cheering crowd, then sent affidavits to Paris, testifying to his success.

The French Academy of Sciences finally decided that the outback never could have succeeded where Paris failed. They denied him a license. A few years later, the French Revolution drove Jouffroy out of France. He died poor and embittered.[11]

All this, however, was still taking place in the eighteenth century—and the subject was still power, not speed. The texture of these attempts began changing in America. Our revolution had been completed, and a different mentality now accompanied invention—the mentality that was ultimately reflected in Emerson's quote about speed.

And here John Fitch, a true American original, enters our story. His work is described fully in a new book by Andrea Sutcliffe.[12] In a nutshell, Fitch was born in 1743 in Connecticut, had a hard childhood, became a silversmith, served for a while as a junior officer in the Colonial Army, and finally went off to the Ohio River basin (ostensibly as a surveyor) to seek his fortune there.

But he was captured by Indians and spent time as a prisoner of the British. He was held in an encampment on a river island near Montreal. In an odd replication of Jouffroy's experience, he gazed at the river traffic. Two years after he was released, a passing remark by a stranger ignited the idea of building steam-powered boats to navigate those sprawling western rivers.

Starting in 1785, Fitch and a competitor, James Rumsey, looked for money to build steamboats. The methodical Rumsey gained the support of George Washington and our new government while Fitch found private backing. He and a German mechanic, Johann Heinrich Voigt, built an engine that combined features of both Newcomen's and Watt's steam engines. They moved from mistake to mistake until, in 1787, they built their first working steamboat.

An 1832 image of Fitch's first boat, from *The Edinburgh Encyclopaedia*.

It was an odd machine, driven by racks of Indian-canoe paddles. The following year they changed the design and kept improving it. Finally, in the summer of 1790, they had a steamboat that ran fairly consistently. It was now powered by a paddle wheel at the stern.

All summer, the boat carried passengers between Philadelphia and Trenton, logging thousands of miles and traveling at about 7 miles per hour. In the end, the effort failed commercially, but that fact should not distract us from the magnitude of the accomplishment.

Fitch was intense, driven, and sometimes erratic, and he had constant trouble sustaining financial backing. But he had shown that a steamboat could provide greater speed than a walking human, and could even compete with the horse over long distances. So if we must name a canonical inventor of the steamboat, perhaps Fitch should wear that crown. After Fitch, it could only be a matter of time before people would be served by regular steamboat service.

The story of the next successful steamboat builder begins with a retired British banker, Patrick Miller.[13] Miller and an employee worked on creating a steam-powered pleasure boat for use at his country estate. He eventually turned for help to the Scottish engineer William Symington. Symington had already built a working steam-powered automobile, and in 1788 he assembled a steamboat for Miller.

He made a second, larger boat a year later. Unloaded, it traveled (like Fitch's boat) at 7 miles per hour. But Miller was very critical of the clumsy rack-and-pinion drive system that Symington had used. He lost interest and went on to other things. The idea lay fallow for another twelve years.

Then Lord Dundas, governor of the Forth and Clyde Canal Company in Scotland, asked Symington if he could build a functional *commercial* steamboat. This time, Symington was cleverer about the drive system. He developed a slider-crank arrangement and in 1803 unveiled a fine steam-powered tugboat. He named it *Charlotte Dundas* after his patron's daughter. That March, on its maiden voyage, the *Charlotte Dundas* carried Lord Dundas (as well as an archbishop) on a six-hour, 19.5-mile journey along the Forth and Clyde Canal.

That may seem slow, but the *Charlotte Dundas* was hauling a pair of seventy-ton barges, and it was doing so against strong headwinds. The voyage was, in fact, a stunning success, and so Lord Dundas ordered eight more steamboats to be built. Then conservative forces rose against a new (and hence fearful) technology. Opponents argued that the paddle wheel action would erode the canal banks. The enterprise finally collapsed.

Still, steamboats were now on the drum. Just the year before Symington's demonstration run, a fateful meeting had taken place between two Americans in Paris. The wealthy promoter Robert Livingston, who had already made an offer to Morey for rights to his steamboat (Chapter 2), was there on business in 1802. He met with Robert Fulton, who was promoting his submarine to the French. Livingston told Fulton that he was prepared to invest substantial money in creating a practical steamboat. At that point, Fulton began moving away from submarines.

Steamboat historian Thomas Flexner portrays Fulton as far "less interested in originality than results."[14] Fulton knew the history of previous steamboats very well, and he made no priority claims to the basic idea. First, he tried to build a steamboat for Napoleon and failed because of the nagging power-to-weight-ratio problem. One boat sank. Another one managed to run, but it was slow.

Before Fulton returned to America in 1806, he had the foresight to have Boulton ship a Watt engine to America—without mentioning steamboats.[15] Now he mounted that engine in a boat, making good use of his bitter experience with sizing boat/engine configurations. On August 17, 1807, he made his famous 150-mile trip from New York City to Albany, moving at about 5 miles per hour. He stopped part way to spend the night at Livingston's home, named Clermont, and finished the run to Albany the next day.

His name for the boat, by the way, was the *North River Steamboat*—not the *Clermont*. Within a few years, Fulton's boats were carrying freight

Model showing the structure of Fulton's *North River Steamboat (Clermont)*(London Science Museum).

and passengers on all the major rivers of America, and the oceangoing steamship would soon follow.

After Fitch, however, no steamboat—not Fulton's nor anyone else's—would ever again compete with land vehicles for speed. Still, although the steamboat did not achieve high speeds, it was nevertheless a primal player in the drama of our quest for speed.

We catch a strong hint of that role in Fitch's competitor James Rumsey, for his boat actually resembled the aeolipile. He did not use his steam engine to drive paddles. Instead, Rumsey picked up on an idea suggested earlier by Benjamin Franklin. His boat was driven by a subsurface water jet.[16] He directed the force of his steam upon one side of his engine's piston. He introduced water he'd scooped up into the cylinder on the other side of the piston. He ejected the resulting high-pressure water through a backward-facing nozzle. His water jet made a perfect enactment of the recoil principle that drove an aeolipile—or a rocket.

Rumsey's arrangement was, in fact, far more radical, and inherently simpler, than the designs that succeeded. But, like the aeolipile, this version of jet propulsion would become efficient only at higher speeds. Another century would have to pass before it would be widely put to use.

Even without Rumsey's jet-powered boat, steamboats would continue to feed the nineteenth-century impulse toward speed out upon the oceans.[17] The first attempt at an Atlantic crossing took place in 1819. Owners of the packet *City of Savannah* mounted engines in their ship, loaded it with coal—no cargo or passengers—and set off for England. They made it there, although they had to use sail for most of the journey.

Thus began a long argument. In 1835, the noted technical handbook writer Dionysius Lardner deduced a theoretical upper limit for the range of a steamboat. He calculated that no ship could carry enough coal to sail more than 2,500 miles. An American lawyer named Junius Smith had been present when the *Savannah* steamed into Liverpool and was sure that Lardner was wrong. He observed that the coal *capacity* of a ship rises as the cube of its length, while coal *burning* rises only as the square. That being the case, all one had to do was to make steamships larger—perhaps larger than ordinary sailing ships.

The daring engineer Isambard Kingdom Brunel also thought it could be done, and in 1837 he began building a large steamer that he named *Great Western*. Smith had by then set up a company, but he had not yet built anything. When he saw that Brunel was about to beat him to the punch, he hastily bought a coastal steam packet, rigged it for the open sea, and renamed it *Sirius*. It was smaller than he would have liked, so he dispensed with any payload. He carried only coal.

In 1838 *Sirius* set out for New York from Bristol. *Great Western* followed out of London three days later, and the race was on. Both were now equipped with more efficient engines than the *Savannah*'s. As for passengers, the *Sirius* carried none, while the *Great Western* carried only seven.

Since *Sirius* was inherently the faster ship, *Great Western* stokers worked night and day, driving its engines at their maximum capacity. *Sirius* hit bad headwinds and ran out of coal. (Could Lardner have been right after all?) When people on Long Island saw her coming, she was burning her supplies for fuel. But she managed to reach New York Harbor and found a cheering crowd awaiting her there.

The *Sirius* had crossed in record time. But eight hours later, *The Great Western* steamed in and bested the record by three days. Neither ship drew many passengers for the return trip to England, and both incurred huge financial losses. But in the wake of that great race everyone began building oceangoing steam packets. Steam had not only conquered the Atlantic but also further whetted our rising appetite for speed.

An oceangoing steamship in an 1832 conception—after *Savannah* but still before *Great Western* (*The Edinburgh Encyclopaedia*).

So ships were still in the game, even though they would never again touch the speed (as measured in raw miles per hour) that could be reached on land. And now it was clear, primarily through what we had seen of steam-powered vessels, that a new high-pressure engine technology had to be created before steam could serve transportation effectively.

Once steam pressures could be pushed to fifty or a hundred pounds per square inch, engines could be made much smaller. And with the evolution of late-eighteenth-century precision boring mills, it was finally becoming possible to make tight-fitting pistons capable of holding high pressures. (Recall the sloppy-fitting Newcomen piston shown in Chapter 5.)

So consider again that word we used for trains when we were children, *choo-choo*—that sound of uncondensed steam exhausting into the atmosphere. It was a signal that an engine was no longer using the condenser, which had once been so necessary to create efficient steam power.

The earliest steam engines had derived power by letting the atmosphere press against a piston while a vacuum was created by steam condensing on the other side. In Watt's engines, steam was introduced under modest pressure, and the work done by the engine came partly from the direct application of steam to a piston and partly from the added boost given by a vacuum in the condenser.

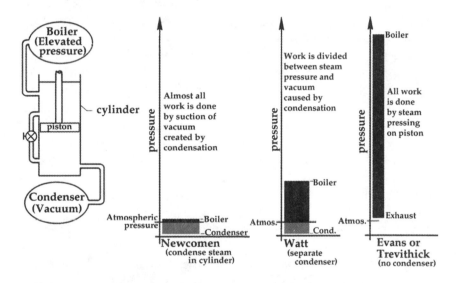

Then, just after 1800, the Cornish mechanic and inventor Richard Trevithick and American millwright Oliver Evans each began making high-pressure engines. That meant smaller cylinders and smaller engines. Raising the steam pressure to several atmospheres reduced the fractional contribution of the condenser to perhaps only 10 percent of the total power output. That much, it seemed, was worth sacrificing to avoid hauling a condenser around in a vehicle.

With the condenser eliminated, steam-powered vehicles finally began making real sense. Instead of condensing steam to create a vacuum, engines simply blew off the spent steam to make that wonderfully nostalgic sound. Indeed, one image that we've all seen in the movies is the water tower alongside the old western train station. Since engines didn't recycle the condensed steam and reuse it, they had to carry water with them, and they frequently had to replace the water that was lost.

In stationary power plants, of course, condensers soon pay for themselves in the saving of fuel, and we have always retained them.[18] (The purpose of those large cooling towers on nuclear power plants is to cool the condensers that receive the steam leaving their turbines.) Even the steamboat *Sirius* had reverted to a condensing engine, both to burn less of its precious fuel and to avoid having to carry a great deal of fresh water.

Evans and Trevithick both pushed steam pressures to at least ten atmospheres. Both used their high-pressure, non-condensing engines to power primitive cars. Trevithick produced a car that could consistently achieve 9 miles per hour. And in 1804 Evans built an amphibious steam-powered dredge, which ran successfully on both land and water. With a proper classical flair, Evans named his machine *Orukter Amphibolos*.[19]

Trevithick, on the other hand, after devoting considerable effort to steam cars for use on roads, turned his attention to providing steam locomotion for rail traffic. Horse-drawn railway systems were already used to move ore in mines, to move industrial goods over modest distances, and as connectors between canal ports. Trevithick began his work on rail systems in 1802, and he demonstrated his first working locomotive in Wales in 1804. It was too heavy for the track, but it made the point nevertheless.

In 1808, he ran a little closed-circuit demonstration railroad in London—a sort of carnival ride with a locomotive called the *Catch-me-who-can*. It moved at a swift 12 miles per hour. And here we discover a syndrome that dogs us all down through the speed-obsessed nineteenth century. Many accounts of the *Catch-me-who-can* run that number up from 12 to 20, on the basis that it really was capable of higher speeds.

We might call the next twenty-one years the incunabula of the railway. A number of British builders, and some Americans as well, began promoting and building rail transport systems. They experimented with all kinds of problems peculiar to this new use of steam—compact boiler designs, optimal configurations of the piston-drive-wheels system, and so forth.

One question in particular had to be resolved: would friction alone provide enough traction between iron wheels and rails? For a while, some builders felt they needed to use rack-and-pinion arrangements in which a toothed wheel engaged a toothed rail. That, of course, proved unnecessary for most use, although rack-and-pinion systems are still used in very steep rail service today. Obviously such a set of necessarily loose-fitting gears was ill suited for speed. Leaving racks and pinions behind, however, appears in part to have been a matter of developing the needed faith in pure iron-to-iron friction.

One railway builder in particular emerged from this new pack of inventors. He was Robert Stephenson, a very able British engineer hired

Detail of the rack-and-pinion drive still in use in New Hampshire on the Mt. Washington cog railway today.

by a developer named Edward Pease to build and equip a twelve-mile-long railway system.[20] Stephenson completed a successful system for carrying both freight and passengers in 1825. It was the first commercial steam-powered railway.

One more step was required, even after this success, before the steam locomotive would finally see full acceptance. Companies still needed assurance that it made sense to use a traveling engine, instead of horses or a fixed engine with some system of towropes. When Stephenson unveiled his *Royal George* engine in 1827, a locomotive far in advance of anything yet, the debate about means for rail locomotion sharpened. Speed was, for the moment, not an issue specific to these debates. After all, a horse could easily have kept up with any of the existing trains. Indeed, these rail systems would still occasionally use horses interchangeably with engines.

The whole matter came to a head in 1829 when the embryonic Liverpool and Manchester Railway sought to resolve the matter of motive power on rails by means of a contest: the Rainhill Trials, with a £500 purse.[21] The trials included tests for hauling capacity, ability to negotiate grade, and speed. Only three entrants made it into the trials: Timothy Hackworth's engine, the *Sans Pareil*; John Braithwaite and Jon Ericsson's *Novelty* (years later, Swedish engineer Ericsson would go on to design the Union ironclad *Monitor* during America's Civil War); and Robert Stephenson's new locomotive, the *Rocket*.

What a name to choose for it! Was Stephenson thinking of celebratory fireworks? Was he thinking of Congreve's war rockets? We can only be certain that he was thinking of *speed*. And if speed had not been much of an issue before, it became one now. In the end, Stephenson's *Rocket* was the only engine to finish the trials, and it was a very clear winner.

Nevertheless, Braithwaite and Ericsson's *Novelty* was a very light and beautiful machine that appears to have been inherently the fastest of the three entrants. It reportedly reached speeds over 30 miles per hour. But here we encounter the daemon of hyperbole that makes it so difficult to know nineteenth-century railroad speeds. A locomotive, especially if there is the slightest downgrade, can go very fast for a moment. When those spurts are combined with wishful thinking, the reports seriously outrun hard data on rail speeds. The hard fact is that *Novelty*'s boiler burst, and it did not finish the trials.

Only the *Rocket* passed the full set of trials, and in doing so, it showed that it could sustain a speed of 24.1 miles per hour. No horse could keep up with that rate for any distance, and in that sense, speed had been achieved. In making the transition from Trevithick to Stephenson, rail had reached its maturity.

Stephenson's *Rocket*. *Left*: As represented by S. Smiles in his 1859 biography of Stephenson's father. *Right*: A scale model in the Boston Museum of Science.

Stephenson and Trevithick were strongly contrasting people. Trevithick came out of the mining regions of Cornwall, gained minimal education, and demonstrated an unparalleled genius for building things. During his life he constructed every kind of steam engine for every kind of purpose. He was tall, strong as an ox, brash, and, as it happened, unlucky in business.[22]

Stephenson was quite another cup of tea. His father, George, was already a successful mechanic and builder of rail equipment who saw to it that Robert got the education he had not received. Son Robert carried himself like a gentleman. He was handsome and a charmer, and where Trevithick might make enemies, he had an instinct for smoothing things over.[23] He built very well indeed upon what others had learned, and he took rail to a new level.

Directly after his success with the Rocket, Stephenson built a still better locomotive. He named it the *Northumbrian,* after the region he had come

Typical steam locomotive firebox on a locomotive, still used on the Silverton Durango Railway.

from. Here, for the first time, he mounted the firebox directly beneath the boiler, as was done in steam locomotives ever after. Although overshadowed by the fame of the *Rocket,* the *Northumbrian* now set the pattern for future locomotive design.

In 1831, the *Northumbrian* made the run between Manchester and Liverpool at about 26 miles per hour. Here again the daemon of hyperbole touches our talk of speed, and it comes from an unexpected quarter. A beautiful young actress, Fanny Kemble, playing in Liverpool at the time, had been invited to make this record-setting train ride with George and Robert Stephenson. Later in her life, Kemble left us with a remarkably detailed account of the trip—a wonderful picture of the visceral excitement that it provided.[24] In it, she attributes a speed of 35 miles per hour to the train. Perhaps it did indeed go that fast in certain downhill stretches of the line.

As a sidelight to her enthusiasm, in his book on rail Nicholas Faith includes a brief section titled "Fanny Kemble in Love." He quotes her claim, published years later, to have been terribly taken not with dapper young Robert Stephenson but with his craggy father, George. "His face is fine, though careworn, and bears an expression of deep thoughtfulness," she writes. "He has certainly turned my head."[25] What exactly she was in love with is unclear, for so much was all rolled together on that day. The intoxication—the excitement—is contagious. Here was movement, speed, personal magnetism, and of course the obvious beginning of a new era in human transportation.

As Robert Stephenson brought steam locomotion to full public service, he assumes a role in the evolution of railroads very similar to Fulton's role in the evolution of the steamboat. Stephenson opened the floodgates of the new transportation technology. After him, rail would spread across Great Britain like ivy across a wall.

America was also in the railway game by the time Stephenson was getting started, and we pushed the technology far more recklessly than Great Britain had. We went to even higher pressures, we laid track with less caution (quickly and over far greater distances), and we invented on the fly as we went along. We reveled in speed.

Still, the odd part about all this is that it was all done with piston engines. The concept of the steam turbine had preceded the piston engine, and turbines held a greater potential for speed. Yet the turbine, for the moment, waited in the wings. The concept of the turbine was more obvious and would ultimately be a lot more effective, especially at high speeds. So what was the catch? Why were piston steam engines built first, and why did they linger so long?

Turbines involved some sophisticated fluid mechanics, but the essential idea of momentum conservation was well understood during the eighteenth century. In fact, Watt, Trevithick, and others had tried and failed at making rotary steam-driven machines.[26] But two additional hurdles faced would-be builders of turbines. To be effective, steam turbine blades would have to incorporate sophisticated aerodynamics, even before airplanes did. And the delicate heavy machining equipment needed to make a turbine was only just coming into being in the early nineteenth century.

The French began creating water turbines during the 1820s, and they were soon followed by other countries—Great Britain and the United States in particular. Like America, France had a rich source of hydroelectric capacity in the French Alps. (By the way, they eventually called their hydroelectric power by the wonderfully evocative term *houille blanche*—white coal.)

Steam turbines would not be commonplace until the American Charles Parsons began producing them in 1887.[27] But an embryonic version of the steam turbine turned up just about the same time as the first water turbines. Frederic Lyman writes about a little-known figure in the creation of the steam turbine, William Avery.[28]

Born in 1793, Avery grew up working as a mechanic in New York State. In 1831, he and a friend were granted a patent for a steam-powered device very similar to those ancient Egyptian toys. A boiler fed steam into the hub of a whirling, propeller-like tube, and jets of steam emerged from each of its tips, causing it to rotate at very high speed. It was our traveler in the forest, Hero's turbine, our sufflator—in yet another manifestation. But the tips of Avery's rotor approached the speed of sound. On one occasion, his cast-iron rotor failed, and fragments tore through three floors of a building.

To get power out at useful speeds, Avery used a series of belts and pulleys. And his engines were put to good use. By 1837, he had built approximately seventy of them. They typically developed around 20 horsepower and were used to drive sawmills, cotton gins, gristmills, and other such small production enterprises.

Among Avery's many engineering accomplishments, he also built what was probably the first steamboat on the Erie Canal. But he died at forty-seven, and the company that manufactured his turbines went bankrupt soon after. We saw little more of steam turbines until the century's end.

Charles Parsons finally made turbines that we would recognize as such, with steam directed through a succession of rotating blades, each stage removing energy and reducing the steam pressure. Fairly complex aerodynamics dictated the shape of the small blades around the rim of the wheels in each successive stage.

As in Avery's turbine, the inherent high speed of rotation would have been an impediment for Parsons. However, the timing was right. Electric generators were just coming into being as well, and they had to rotate as relatively high speeds. The turbine was a generator's natural mate—the perfect means for putting its power to use.[29]

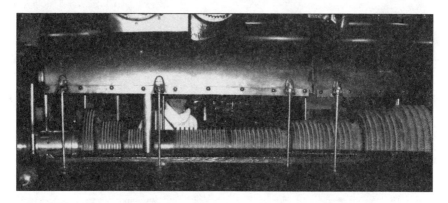

A cutaway 1891 Parsons axial-flow turbine, revealing a series of dozens of stages of turbine blades (London Science Museum).

The steam turbine would be followed by the gas turbine. Then the gas turbine would be united with principle of jet propulsion to give us the airplane jet engine. It would become the agent of speed in the twentieth century, when speeds would become higher but gradually begin to mean less. But the relevance of the steam turbine to the nineteenth-century fixation on speed was made clear enough in 1894. The hastily formed

Turbinia—the turbine-powered launch.

Marine Steam Turbine Company built the 100-foot, turbine-driven *Turbinia*. It reached what was then an impressive speed on water—33 knots or 39 mph.

But all this was the stuff of a new century. The archetype of nineteenth-century speed was the locomotive. It had been so when Turner painted *Rain, Steam, and Speed*, and it would be so at the century's end. Walt Whitman recognized that fact with great clarity. In 1855, he began writing his great song to America, *Leaves of Grass*. He celebrated all the new technologies, but none so eloquently as the railways. In 1876, he added a poem titled "To a Locomotive in Winter." "Type of the modern," he calls the great new locomotives, "emblem of motion and power—pulse of the continent." He finishes with these lines, in which we can fairly taste the manic momentum that was now driving us:

> Law of thyself complete, thine own track firmly holding,
> . . .
> Thy trills of shrieks by rocks and hills return'd,
> Launch'd o'er the prairies wide, across the lakes,
> To the free skies unpent and glad and strong.[30]

There now seemed no limit to speed. Rail speeds continued rising. After Stephenson, records of actual locomotive speeds are almost all generally flawed in one way or another by the same hope and excitement that touched Fanny Kemble. Hold any nineteenth-century locomotive speed

Empire State Express No. 999 (Chicago Museum of Science and Industry).

record up to the light, and you will find ways in which it was probably fudged.

For example, on May 10, 1893, in conjunction with the Chicago World's Columbian Exposition, the New York Central and Hudson River railway did a trial run of their *Empire State Express No. 999* in upstate New York. Recorders on board clocked it at 81.5 mph, and that is all that the railway (which owned it) claimed. Unofficial timers, however, claimed to have clocked it at 112.5 mph, and that is the record generally quoted.[31]

No matter the many exaggerations. Great speed had been accomplished on rails, even if the mantle was about to pass to other vehicles— to steam-powered automobiles, then to internal combustion machines, and finally (after World War I) to airplanes.

We invented many transportation technologies during the nineteenth century, but the gnawing, underlying motivation for all those machines was *speed*. Therefore, before we go further with our considerations of the early evolution of inventions, we need to better define the attribute of motivation. We need to see just what happens when any invented quality—speed, efficiency, accuracy—begins growing in the way that speed grew in the nineteenth century.

8

Inventive Motivation and Exponential Change

I first heard the old story of the philosopher and the three workmen in the autumn of 1951. You know the one: A philosopher passes three workers who are cleaning mortar off old bricks. He asks each in turn what he's doing, and gets three very different answers.

First worker: "I'm removing mortar from bricks, so we can reuse them."
Second worker: "I'm building a great cathedral here."
Third worker: "I'm seeking to bring people closer to God."

The story was still on my mind the next day at work. I was working on the new B-52A bomber at the Boeing Airplane Company. I stopped by a friend's drawing board in the vast hall of drawing boards where he and I labored on the bomber's pneumatics system. When I told him the story, he looked down at the layout drawing he was working on. He was trying to fit a high-pressure air duct in the bomb bay without entering the large space reserved for bombs. He shook his head sadly and said, "I'm seeking to bring people closer to God."

I suspect that, unless we are in some way morally deformed, each of us gauges the purpose of our work against different measures at different times. Sometimes we use a micrometer, sometimes a yardstick, and sometimes (if only occasionally) we try to place our efforts within the compass of some ultimate cosmic global positioning system.

One cannot look seriously at any effort, including invention, without asking about motivation. The question "Why do we invent?" becomes so puzzling because the inventive process plays tag with all the stimuli that drive us this way and that. Let us try some of the more obvious stimuli.

Profit. We all seek, to one extent or another, after personal gain. We all want things. Those things can vary all the way from basic food and warmth to palaces and yachts. If invention promises to serve our wants, and if we believe in our ability to invent, then we will use our inventions to secure profit. Thus, if we place our ability to invent in the service of profit, that hardly means profit is motivating invention. That would be a bit like saying that listening to the radio is our motive for buying a car, because we like to listen when we drive.

Necessity is merely a variation on profit. We satisfy certain needs because we have to. One might add the related motivation of controlling nature. If invention is a tool that we are able to bring to bear upon our problems, then we will certainly do so. I once ran away from a man trying to hold me up at gunpoint. However, I had learned to run not so that I could avoid muggers but because running gave me physical pleasure.

So let us consider *pleasure*. If we have ever invented, and we all have, we need only reflect upon the pleasure it gave us to do so. If we have ever seriously bitten into the superb pleasure of invention (or of such corollaries as discovery or fresh learning), we have found ourselves wanting to do so again. Whether or not Archimedes really ran down the road naked shouting, "Eureka," the story survives because we all understand why he *might* have done so.

And there is *freedom*. Inventing means violating some status quo. If we do not exert some freedom of rebellion, we do not invent. It might be freedom from external proscription, or it might be freedom from chains forged in our own minds. All the great inventive epochs of the world have been marked by climates of increased personal liberty. In Chapter 4 we describe that climate in the Hellenistic world. We see, in Chapter 5, such a period opening up during the eighteenth century. There have been many other such periods throughout history. We could tell a similar story about China's early Sung Dynasty (AD 960–1127). If we want to stimulate invention, then we need to somehow create an ambience of freedom. If you and I hope to access invention, then we need to find that ambience within ourselves.

Unfortunately, these various motivations are like apples and oranges; they don't belong in the same list. The forces of profit or necessity can stimulate *any* use of our talents—invention, hard work, deceit, et cetera. Both can be used in the service of profit or of necessity. Freedom, on the other hand, while it is conducive to invention, does not aim invention toward anything. People driven by need or exigency often, therefore, try to curtail and control the inventor's freedom out of their own sense of urgency. The result, of course, is the same as cutting open the golden goose to retrieve all her eggs at once.

But pleasure is unique among these drivers because it is the single sure reward of successful invention. Like freedom, it offers no goal of its own. Rather, it aligns itself with other pleasures. Flying machines are an example. They offered a kinesthetic pleasure that so many people wanted to access. Above all else, invention is hedonistic. Inventors want the buzz of having created something good and new. They also want the buzz that this something might offer.

Consider, for example, the flowering of powered transportation in the nineteenth century. Some people worked on steamboats, some on locomotives, others on automobiles. Some inventors worked primarily on engines, others on how power might best be used to create motion. So what motivated them?

They generally fought for profits, and they all argued that they were fulfilling necessity. But we also find that each of their personal stories expresses a love for freedom—freedom of movement, freedom of the mind. Indeed, most of them freely switched, without any sense of betrayal of purpose, from one technology to another—from engine building to steamboats to cars to locomotives. And finally, they all gained great pleasure from their successes while they generally suffered failure badly.

Yet one overriding purpose seems to have touched them all. I argue in Chapter 7 that they all held a mental picture of speed in their heads. What appears to have touched the late-eighteenth-century psyche was the realization that the age-old craving for speed might be attainable now that we had steam engines. There was a Zeitgeist, a collective unconscious, and its name was speed.

Let us then trace the evolution of speed as measured in miles per hour. Once we have done so, we shall be able to use it as a springboard from which to consider the character of the exponential change so often observed in various technologies.

We noted several important speed records in Chapter 7: Fitch's steamboat, Trevithick's car and his locomotive, Stephenson's locomotives, the *Empire State Express,* the *Stanley Steamer.* These are all combined with additional data, including the later record of internal-combustion-powered vehicles, in Table 8.1.[1]

When we plot these speeds on semi-logarithmic coordinates, each against the date it was achieved, they suggest an almost perfectly a straight line. What that means is that the maximum speed anyone had reached by any date rises exponentially. We can represent these data with the equation

$$\frac{speed}{8.09\ mph} = e^{\frac{date-1790}{41.2\ years}}$$

$$(1)$$

Since the exponential increase is so consistent (the correlation coefficient is a very high 0.994) we immediately have to wonder if this growth behavior is typical and if we can explain it.

One way to answer these questions is to create a model for technological change and then compare it with data for additional technologies.[2]

Table 8.1

Speeds at the Earth's Surface

Class and Specific Machine	Date	Speed (mph)
Steamboat		
Fitch's steamboat	1790	7
(After Fitch, steamboats never again moved faster than land vehicles)		
Steam Cars		
Trevithick's car	1803	9
Stanley Steamer	1906	127.7
Stanley Steamer	1907	150
Railroad		
Trevithick's locomotive	1808	12
Stephenson's *Rocket*	1829	24
The Northumbrian	1830	26
Brunel's *North Star*	1838	33.5
Empire State Express 999	1893	82
Siemens and Halse electric	1903	130.6
(After this, railroads never again caught up with automobiles)		
Internal combustion automobiles		
Barney Oldfield	1910	131.5
Henry Seagrave's *Golden Arrow*	1927	241.4
M. Campbell's *Blue Bird*	1935	301.1
John Cobb's *Railton*	1939	369.7
Craig Breedlove's *Spirit of America*	1963	408.3
Spirit of America—Sonic 1	1965	600.6

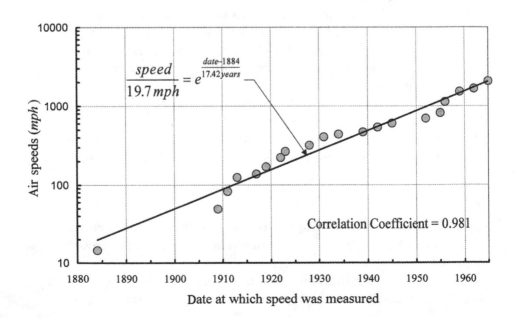

$$\frac{speed}{19.7\,mph} = e^{\frac{date-1884}{17.42\,years}}$$

Correlation Coefficient = 0.981

Date at which speed was measured

If the data match the model, then chances are good that the model is correct. To be certain, however, we need more than a simple data match. We need convincing reasons why growth should occur in this way as well.

We begin by imagining some quality (call it Q) that gauges the improvement of a technology. Speed is one example. Others might be efficiency, clock accuracy, or anything else that inventors work to improve. We do have to impose some limitations on Q, however.

- The quality obviously has to be *quantifiable.* We cannot do much with a desirable but immeasurable quality such as, say, beauty. Speed makes a very useful candidate since it is a clear numerical measure of something that we prize in transportation systems.
- The technology must be *complex* enough that it would depend upon many inventions. We have already looked closely enough at the quality of speed to know that systems for giving us rapid motion have been very complex. No one inventor could have given us steamboats, steam engines, railways, or airplanes. We have seen that even the lowly doughnut represents a great deal of individual invention.
- The technology must be *ingenuity-limited.* By that we mean it must still be within the realm of human ingenuity to improve it. It must not have run up against some physical wall. To pick two extreme cases, the accuracy of clocks is ultimately limited by Heisenberg's uncertainty calculation—no time can ever be stipulated with absolute precision. Similarly, speeds are ultimately bounded by the speed of light. Ingenuity will not get us around either of those hurdles.

Although the speed of land transport is subject to much more modest limits than the speed of light, none of the data in Table 8.1 has reached those limits either. The data we have looked at represent stages achieved by the untrammeled creative abilities of successive inventors.

The efficiency of modern engine-driven power plants is a quality that is clearly no longer ingenuity-limited. The laws of thermodynamics (Chapter 6) tell us that the efficiency can never reach 100 percent. As engines now approach 50 percent efficiency, it is clear that ingenuity will never again double their thermal efficiency.

- The advance of the technology must be *motivated.* This is the key issue in our study of invention. By *motivated,* we mean that there is widespread agreement that a technology needs to be pursued. Speed became a good example only after we had perceived that we finally had the means (vehicular steam engines) for accomplishing it.

As a counterexample, consider the height of tall buildings—a quality that has never sustained exponential increase. Ever since Babel or the lighthouse at Alexandria, the quality of height has begat occasional obsessive bursts of effort. A few late-medieval cathedrals reached dizzying heights, and early-twentieth-century skyscrapers far outdid even them. Every now and then, some city decides to build a grander and taller structure than any other city has.

But the quality of height has never been motivated in the sense we wish to use the word. For one thing, height is not entirely functional. It does save real estate, but it also poses formidable problems of parking, cost, and vulnerability to wind or attack. As buildings become very tall, the fraction of the plan area that must be committed to elevators becomes large. And a very tall building can have the effect of disconnecting the people within it from one another. While height might gain motivation in some localized sense, it has never done so in the collective way that leads to exponential change.

The period during which a technology is motivated and ingenuity-limited usually ends in either of two ways:

- The technology might be *completed*. Completion is one of two ways in which the ingenuity-limited period might end. We call a technology complete when it reaches some physical limit. The efficiency of engines, for example, has long been complete because, being limited by thermodynamic laws, it can no longer be radically improved by any human agency.
- A motivated and ingenuity-limited technology can be also be *replaced* with a wholly different technology if such a means better serves its function. The accuracy of mechanical clocks, for example, improved for six centuries. Then, in the early twentieth century, it began crowding not quantum limits but the more pedestrian limits of machine making. However, around 1920, mechanical clocks were *replaced,* first by electrical timing elements, then by quartz crystals. Since then, the technology of accurate timekeeping has become a wholly new enterprise.

Once we have these definitions, we introduce our major assumption—a key hypothesis that will have to be verified after the fact. It goes like this: *The people who develop and use a particular technology tacitly settle upon a common expectation. It is this: from generation to generation they will continue to improve their technology by a certain factor during their working lifetime (like doubling it, or quadrupling it).*

Although nothing in the model depends upon it, it helps fix our thinking if we imagine a typical length for a working lifetime. We shall take it to be thirty years, bearing in mind that the period of active practice varies widely around any such value.

Now, to continue we need just a little arithmetic. We note that doubling means creating a twofold increase every thirty years; quadrupling would be raising it fourfold. Generally speaking, we might have an n-fold increase, where n is two for doubling, three for tripling, and so forth. If we are driven, say, to double some technological quality, Q, every thirty years, then we can express that increase arithmetically as follows:

$$\frac{Q}{Q_0} = 2^{\frac{t-t_0}{30}} \tag{2}$$

where the symbol t designates time and the subscript o identifies the starting date and quality. This says that, after sixty years, Q will have doubled twice and will be four times as large, and so forth. For any other n-folding, we would have:

$$\frac{Q}{Q_0} = n^{\frac{t-t_0}{30}} \tag{3}$$

The more conventional way of representing exponential behavior is in terms of the number $e = 2.71828 \ldots$ (the base for natural logarithms) and a so-called time constant, T:

$$\frac{Q}{Q_0} = e^{\frac{t-t_0}{T}} \tag{4}$$

If we take the logarithms of both sides of equations (3) and (4) and compare their right-hand sides, we find that the time constant depends on the n-folding rate in the following way:

$$T = \frac{30}{ln(n)} \tag{5}$$

where the symbol ln signifies a natural logarithm. Thus, a more rapid increase of a quality means a greater n-folding but a smaller time constant.

Normally, an engineer would deal only in time constants and not bother with n and n-folding. But relating n to the notion of a technological lifetime offers evocative means for creating a mental picture of the subjective substance of technological change. Considering the increase of speed at the earth's surface, starting with Fitch's 1790 steamboat, we get a time constant of 41.2 years. Equation (5) then tells us that everyone who worked with those transportation systems in the nineteenth and twentieth centuries looked for an n-folding of 2.06—they expected to see speeds roughly double.

This curve ends with Breedlove's 601 mph run on the Bonneville Salt Flats. At that point he was crowding the speed of sound. He was now subject to such aerodynamic concerns as keeping the vehicle in contact with the ground. These matters were complicated by the odd character of supersonic aerodynamics, to which the vehicle was now subject. The British *Thrust SSC* achieved a Mach number of 1.02 in 1997, but it was jet-propelled (like the aeolipile in Chapter 7), no longer wheel-driven. Its characteristics were more those of a wingless supersonic airplane than an automobile, and growth of the quality of speed was essentially complete.

Land transportation speeds, while they represent many branches of technological change, yield only one exponential curve. Let us then move on to other technologies. Consider the efficiency of steam engines. Here we have a problem since, up to the mid-nineteenth century, the principle of energy conservation was still unknown, and there was no concept of efficiency as a ratio of work done to heat supplied.

Early steam engineers did, however, use a quality called the *duty* of an engine. An engine's duty was the number of foot-pounds of work it could extract from a bushel of coal. We are fortunate that the information needed to calculate a thermal efficiency is embedded in the duty. We can obtain the efficiency from it if we know what kind of coal was used and how the bushel was defined in the unsettled weights and measures of the time.

The eighteenth-century engineer who introduced the term *duty* was John Smeaton—a true engineering polymath. Smeaton designed the first Eddystone lighthouse that was able to endure in the fierce winds and waves around it, he made important studies of waterwheel design, and he worked extensively on Newcomen's engine, greatly improving its performance.

The earliest value of steam engine duty for which we have any record is Smeaton's measurement for an existing Newcomen engine. He got 5.59 million ft-lb/bushel.[3] We can estimate that the coal's heating value was 13,500 Btu/lb, and it is likely that he used 84-pound bushels. On that basis, the efficiency would have been just under 2/3 of 1 percent (a pitiful figure, by our lights). We assume the engine was built sometime between 1718, when fully evolved Newcomen engines began appearing, and 1767, when Smeaton made his tests. If we average those two dates, then our graph begins with an efficiency of 0.64 percent in the year 1742 ± 14 years.

By 1775, Smeaton had improved upon Newcomen until he had an engine with an efficiency of 1.08 percent. Watt subsequently kept improving steam engines until their efficiency was almost 4 percent. The curve we obtain by gathering such data once again reveals a clear exponential increase of steam engine efficiencies. The time constant is 32.6 years and n is 2.5. Steam engine efficiencies thus rose a bit more rapidly than the quality of speed did; however, it still roughly doubled every thirty years.

We find a clear example of completeness in this case. The curve races upward to the point at which human inventiveness is thwarted by im-

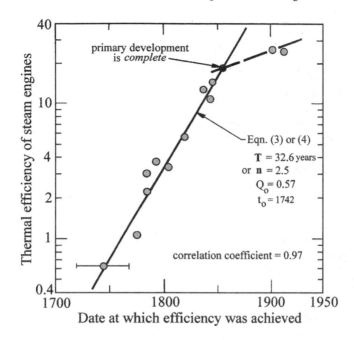

Date at which efficiency was achieved

mutable thermodynamic laws. After around 1850, efficiencies could only creep upward. Today, we pat ourselves on the back when we improve efficiency by 1 percent. Now and then, improved metallurgy allows us to nudge boiler temperatures and pressures upward. Better bearings and seals might gain a fraction of a percent here or there. Efficiencies have risen by a scant 15 percent since the 1930s, while the motivation for improvement has continued to be intense.

Returning to our theme of speed, let us see what happened when powered transportation took to the air.[4] Inventors had clearly been looking to the sky for some time when Henri Giffard flew a navigable engine-powered dirigible over Paris in 1852. It moved at roughly the speed of a walk. Thus, while it is a large landmark in the history of flight, it does not represent the beginning of motivated growth of the quality of speed.

The dirigible that first entered the race for speed was the one developed by Charles Renard. In 1884, Renard and another French army officer, Arthur Krebs, made a five-mile trip in Renard's airship, *La France*.[5] Renard's work was meticulous and included some surprisingly cutting-edge technology. He used a very early battery-powered electric motor to drive his airship, which moved along at a brisk 14.5 mph. *La France* marked the beginning of serious dirigible building.

Table 8.2 lists speed records from that year forward until the 2,094 mph flight of the McDonnell *Voodoo* jet airplane in 1965. The *Voodoo* record has subsequently been bested, but the intense focus (motivation) on the speed of air-breathing airplanes was over.

Table 8.2

Speeds in the Air

Class and Specific Machine	Date	Speed (mph)
Lighter-than-air flight		
La France dirigible	1884	14.5
(After this, dirigibles were outrun by		
heavier-than-air craft)		
Heavier-than-air flight		
(limited to air-breathing engines)		
Bleriot XII	1909	49.2
Bleriot XI	1911	82.7
Deperdussin	1913	124.8
SPAD VIII and S.E. 5A	1917	138
Nieuport	1919	170
Curtiss R-6	1922	223
Curtiss R2C-1	1923	266.6
Macchi M.52bis	1928	318.6
Supermarine S.6B	1931	407.5
Macchi MC72	1934	440.7
Messerschmitt Me209 V1	1939	469.2
Messerschmitt 262 jet	1942	540
Gloster *Meteor*	1945	606
North American F-86D Sabre	1952	698.5
North American F-100C Super Sabre	1955	822.1
Convair F-106A	1956	1,132.1
Mikovan E-166	1959	1,525.9
Fairey Delta Two	1962	1,612
McDonnell F-101A Voodoo	1965	2,094

Even before 1965, several *rocket*-powered experimental aircraft had already traveled faster, but they were kin to the new space vehicles. Four years before the *Voodoo* record, Yuri Gagarin had orbited the earth, and we now saw humans moving at speeds on the order of 17,000 mph. When, in 1959, the North American X-15A, powered by Thiokol rocket engines, flew at 4,159 mph, speed and invention had been decoupled. For the moment, speed was unlimited, and the airplane had been (in our special vocabulary) *replaced*. We turned to other technologies; we began tuning our ears to the murmurings of other Zeitgeists.

A plot of these data reveals yet another clear example of exponential growth. In this case, the time constant is only 17.42 years and the *n*-folding rate is up to 6.00. Flight started out at speeds far below those of railways and was still much slower than automobiles at the beginning of World War I. When speeds achieved by airplanes finally caught up with automobiles in the 1920s, they were traveling at speeds in excess of 200 mph.

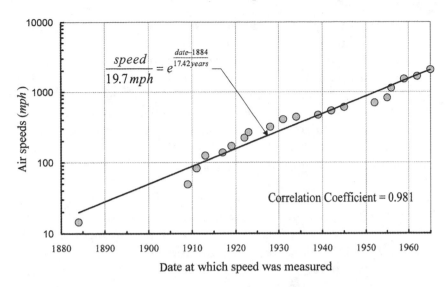

$$\frac{speed}{19.7\,mph} = e^{\frac{date-1884}{17.42\,years}}$$

Correlation Coefficient = 0.981

Air speeds (*mph*)

Date at which speed was measured

The airspeed graph reveals something of the role of necessity as a motivation for invention. Notice how data oscillate slightly about the line they define. The rate of increase is higher during peacetime and lower during wars. The need for higher airspeeds and the tactical advantage that they offer is intense during war. But speeds nevertheless rose less rapidly during World Wars I and II and the Korean War.

While that flies in the teeth of the popular belief that the necessities of war drive invention, it also makes perfect sense. Too much urgency distracts inventors from their goal; you might say it jiggles their aim. Urgency makes it harder for inventors to find the elbow room—the freedom—that invention requires. The abatement, however, is only slight, since the inventive urge is strong and it continues, inexorably, through war and peace alike, despite the dampening climate of war.

That is not to say that other aspects of technology do not thrive during wartime. We witnessed outright miracles of production and adaptation during World War II, for example. But production can be forced in ways that invention cannot.

Another feature of this airspeed curve is that it reveals a great increase in the rate at which speed was advanced by airship and airplane inventors over the rate that it was advanced by automobile and railway inventors. So let us take stock of what we know about those two breeds of technologists.

Late-nineteenth-century rail and automobile pioneers were not interchangeable with pioneers of flight. The two types seldom moved back and forth between these two very different kinds of vehicle. Early automobile makers had often begun working with railway systems or as carriage makers. On the other hand, the kinship between airplane and bicycle makers runs very deep. It turns out that we can see why if we merely count the wheels.

A railway car or an automobile generally runs on at least a four-point suspension. It is inherently stable and solidly bound to the earth. A bicycle is quite another matter. Never stable, it requires constant rider intervention to stay upright. Early automobile and bicycle builders held radically different, and essentially contradictory, views of motion. The Wright brothers, themselves former bicycle makers, achieved their success in large measure just because they understood that principle of motion.

Human bipedal locomotion, like the flight of birds, embodies that idea of instability. We and they live in constant states of unstable motion. If we did not correct our walking, or birds their flight, at every instant, we and they alike would fall. It was through their deep-seated, gut-level understanding of instability that the Wright brothers could succeed where so many others had failed.

Perhaps at a deeper level, they understood that the unremitting correction of constant instability is a defining characteristic of any living being. It has often been pointed out that for any creature to live, it must employ constant negative feedback: it can never stop sensing error and feeding back a negative (or corrective) action. The essence of human freedom is that we are at liberty to err and correct in that way.

The Wrights were only the most notable of the many bicycle makers who so naturally segued into the sky—Glenn Curtiss was a motorcycle pioneer, the French helicopter pioneer Cornu was a bicycle maker, and so on. We also find a most instructive counterexample in automobile pioneer Daimler, who began by making a motorcycle without understanding the interwoven concepts of instability and freedom that underlay two-wheeled vehicles. He put a small pair of stabilizing wheels on either side, creating what can only be called a frightful kludge. Try to imagine a motorcycle with training wheels! Soon after, Daimler joined with Benz to develop the early four-wheeled automobile and gave no more thought to motorcycles.

Automobile makers remained creatures of the late eighteenth century, while fliers helped to define the embryonic twentieth century. For all practical purposes Henry Ford and

Butler's 1885 internal-combustion-powered tricycle illustrates the initial blurring between bicycle and automobile pioneers. That blurring would not last. From the 1911 *Encyclopaedia Britannica*.

Table 8.3

Time Constants and n-Folding Rates

Technological quality, Q	Range of motivated growth—Q_o to date of completion	n	T (years)
Examples developed herein			
Water and land speeds (mph)	1790–*ca.* 1965	2.06	41.2
Steam engine efficiency (percent)	1742–*ca.* 1850	2.51	32.6
Air speeds (mph)	1884–1965	6.00	17.4
Examples previously developed by Lienhard			
Corn yields (bushels per acre). This is an ancient technology that was restarted around 1940 by the new sciences of genetics and biochemistry.	1930	2.68	30.4
Mechanical clock accuracy (days of operation per second of error)	*ca.* 1400–1920	1.95	45.1
Land vehicle engine (power/weight ratio)	1803–1984	2.49	32.9
Well drilling (depth in feet)	1841–1979	2.52	32.5
Diving vehicles (depth in feet)	1865–1979	3.46	24.2
Low temperatures (refrigeration) (1/temp. − 1/298) K^{-1}	1860–1936	21.4	9.8
Incandescent bulbs (lumens/watts)	1880–1979	6.45	16.1
Airplane engine (power-to-weight ratio)	1903–1984	12.7	11.8
Liquid-fueled rocket altitude (ft.)	1926–1956	785.8	4.5
Computers (bits per add time)	1945–1979	7×10^{11}	1.1
Printed circuits (components per circuit)	1959–1984	2.02×10^9	1.4
Examples developed by Struble			
Optical telescopes (aperture/resolution)	1610–1991	1.96	44.4
Radio astronomy (baseline area/wavelength)	1933–1991	1.4×10^6	2.12
Nuclear fusion (temp. × density × time)	1968–1991	2.25×10^8	1.56
Maximum superconducting temperature	1986–1991	8.05×10^{57}	0.23

Glenn Curtiss might have come from different planets. One railway engineer, Octave Chanute (Chapter 3), made the shift, but only after he had retired from rail. And even then, he himself did not try to fly. The two groups of inventors who took up the pursuit of speed did it in very different ways, with the airplane inventors operating in a far more unrestrained way. They created a new technological culture that perpetuated itself until Sputnik, and they begat far more rapid improvements in speed than inventors who had remained down on the ground.

Here then are three case histories: speed on water and land, speed in the air, and steam engine efficiency. Table 8.3 summarizes these and fifteen additional such case histories, taken from the earlier papers on exponential growth rates.[6] They represent a wide variety of technologies over a very long time. Our earlier suggestion that the quality of a technology might double in a technologist's lifetime seems valid for the older technologies, but for the more recent ones, that idea deteriorates in a way that is truly dazzling.

When we plot these time constants against the dates when motivated growth appears to have begun, we find a very clear break around 1841. Up until that time, the average time constant is thirty-seven years, which corresponds to an n-folding of 2.25 in a thirty-year working lifetime. That is very close to a simple doubling, as we had anticipated.

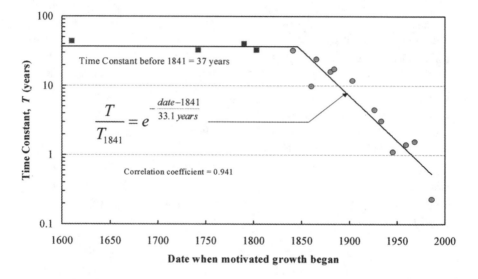

But, after 1841, the time-constant starts to shrink, and it does so exponentially. For over a century and a half, we can identify a secondary time constant that specifies the rate of decrease of the growth time constant. That secondary time constant is 33.1 years—very nearly the same as the average growth time-constant, T, before 1841. Something very strange appears to be going on here, but it can be sorted out by looking at the historical record of the period.

Until approximately 1841, all of the artisans who worked on a given technology, along with the public that they served, tacitly agreed to improve the technology by a factor of roughly two during their working lifetime. It seems they all expected their technologies to improve at that rate, and that is what happened.

Then, around the mid-nineteenth century, that expectation appears to shift very suddenly. A completely new expectation seems to form—an expectation that as each new technology appears, it will improve faster than any before it. After that shift, time constants for the improvement of new technologies are halved roughly every thirty years.[7]

This shift has not gone unnoticed. Historians have suspected it, and pointed it out in qualitative terms. British historian D. S. L. Cardwell, in particular, wrote:

> The conspicuous feature of the period was the rapid convergence of science and technology considered as institutions, the virtual conclusion of the Baconian programme. Whitehead, whose life spanned this period, expressed the point very succinctly when he wrote: "The greatest invention of the 19th century was the invention of the method of inventions."

Cardwell went on to clarify Whitehead's odd phrase "the invention of the method of inventions." It did not refer to some simple formulaic process. Rather, it acknowledged the rise of previously unknown institutions and agents that created new expectations for invention—put invention in the hands of new breeds of people. Cardwell identifies

> the research laboratory, staffed by professional scientists, the design/development department and technical sales and services. What had been, in the past, the highly distinctive features of occasional firms such as Boulton and Watt . . . now became increasingly commonplace in countries with advanced technologies. The main agents of this change have been the increasingly numerous classes of highly trained professional engineers and scientists.[8]

We in America often point to Thomas Edison as the canonical inventor of the R&D laboratory. But, once again, we bury the past when we do so. Cardwell correctly credits Boulton and Watt with having a good sense of the modern R&D enterprise. More significantly, just after Watt's death a young German chemist cast that model into what proved to be very close to its finished Edisonian form. He was the Baron Justis von Liebig, born in Darmstadt in 1803.[9]

Von Liebig took an interest in chemistry when he was seventeen, and at twenty he went to Paris for a year to study with Joseph Louis Gay-Lussac. Gay-Lussac opened his eyes to the new idea that we would need accurate experiments to make sense of chemistry. Von Liebig came back to a post at the University of Giessen, where he turned his young man's enthusiasm, along with remarkable foresight, on Gay-Lussac's ideas. He worked single-mindedly to set up a chemical research laboratory, spending his own salary on equipment. By 1827, he had a twenty-man operation, poised to undertake creative development in a more systematic and inexorable fashion than anyone before him.

One room of von Liebig's laboratory at the University of Giessen. Nineteenth-century print, source unknown.

We honor von Liebig for his work in organic, agricultural, and pharmacological chemistry, but this laboratory was his greatest contribution. Other chemists had to copy it to keep up with him. It formed a watershed in the way we would think about invention. Others—Boulton and Watt, for example—had shaped the beginnings of such an institutional structure. Now von Liebig brought that structure to full bloom.

In 1843, one of von Liebig's former students sent him an oil he had isolated from coal tar. The laboratory found a compound in it that reacted with nitric acid to make brilliant blue, yellow, and scarlet coloring agents. Liebig had already anticipated the compound—a form of benzene in which one hydrogen atom was replaced with an amino group. They named the chemical *aniline,* and by 1860 Germany had built a new dye industry, beginning with aniline. That dye industry, in turn, helped carry Germany into world leadership in industrial chemistry—leadership that owed much to Liebig's ideas about chemistry, but far more to his vision of systematic research, invention, and development.

aniline molecule

Cardwell and other older historians refer to this shift as the beginning of *professionalism*—an interpretation that has more recently fallen out of fashion because the shift had dimensions not accurately contained in that word.[10] (Indeed, I would challenge anyone to give me a mutually agreeable definition of the word *professional.*) It is probably better to identify what happened in the third and fourth decades of the nineteenth century as the development of new institutions for both learning and practicing science and technology.

In any case, to give von Liebig full credit for this shift in vision would be no more right than to credit Fulton, or even Fitch, as primary inventor of the steamboat. The concept that people specifically educated in science could work together to achieve their objectives at an accelerated rate was now on the drum, and many were listening to the whisperings of that Zeitgeist.

Using universities to teach people to be engineers was another new idea in ferment at this time. Both France and England had begun embryonic military engineering schools in the late eighteenth century. These so-called polytechnic institutions were gaining enormous momentum in France and America at precisely the same time von Liebig was building his laboratory.[11]

The shift that took place around 1841 thus reflected a new corporate/communal vision. Once the shift had occurred, each new technology would unfold more rapidly than the one before it. If we look again at the equation for the change of T during this time, we recognize that it is the so-called *breeding equation*—the same equation that dictates the growth of a living population that has not yet been limited (or completed) by such outside factors as living space and food supply.

The equation tells us that, for the last century and a half, the rate of acceleration of technological improvement (the reduction of T) has depended directly on the present rate—just as a population always grows in proportion to its present size. Now, however, it is the rate of technological improvement that breeds in proportion to itself. This is what we call self-excitation, and self-excitation is what Cardwell and Whitehead were describing as well.

We cannot look at rapidly accelerating rates of exponential growth without mentioning the often-quoted Moore's law. Moore's law says that the number of transistors per integrated circuit doubles each year. That translates into a time constant of 1.44. But our equation for decreasing time constants gives 1.3 for the year 1952, when the concept of the integrated circuit was first suggested.[12] Thus Moore's law fits right in with the cases the equation is built on.

Moore's law, by the way, is tantalizing because we have anticipated its completion for some time now. The density of transistors is drawing remarkably close to the molecular limitations of matter, and will have to reach that limit—just as steam engine efficiencies reached theirs. Yet, at this writing, it astonishes us all by continuing its exponential increase.

In any case, motivation touches our major technologies for a season, and when it does, they flourish with a glorious animal insistence. Once it has been caught up in our communal agreement that it is important, a technology grows like a tree root against cement. The individual texture of invention that makes it up is relentless, and it will not be denied.

If those of us working in the technologies lose sight of that fact, we do so at our peril. A motivated technology is a wild horse that we might be able to herd, but whose motion we can neither slow nor urge ahead. Nor can we bridle the motivation that drives it. My repeated use of such terms as *Zeitgeist* and *collective unconscious* reflects my attempt to capture the force driving the technological motivation that stands out so clearly in the various rate curves.

Now let us move on to a completely different set of technologies. These are seated very close to our jugular vein, and they have, down through the millennia, proved to be powerfully motivated. We turn next to the technologies by which we extend the ability to make our thoughts known to one another.

Part III

Writing and Showing

9

Inventing Gutenberg

It may be true that not all invention takes place within the framework of communal motivation that we have been describing. But the test of any invention's importance is whether or not others have built upon it. And if an invention does not reflect a groundswell of collective motivation—no matter how clever that invention is or how great its potential—it will either die upon the vine or have to wait in the wings until it reveals the potential for serving such a groundswell.

We need to keep that in mind when we read about invention, for it is so often portrayed as solitary. Invention is, without a doubt, an individual act. But a new thing can neither come into being nor live, breathe, and grow if a framework of people and artifacts does not buoy it up. To understand the history of any successful invention we therefore need to look first for the mutual craving that drove it. That alone is what provided the invention with a supportive community; that alone is what made its emergence possible.

Thus far, we have parsed several inventions in terms of elemental desires. Those that drove flight and speed were more dramatic, while power production reflected another kind of desire, namely, creature comfort. The desire for comfort emerges in so many ways. Vast invention has stemmed from our craving to be warm, well lit, fed, clothed, and healed of illness.

I must stress the subtle line that separates desire from necessity. Consider what can happen when we need nutrients but desire sweets. Necessity is deceptive. The low-income person who needs money for the rent and instead buys a new car is a familiar enough figure. Desire trumps necessity again and again. If it did not, then invention would generally be dismissed as an impractical activity.

It is important to keep that in mind when we come to another kind of deep-seated desire—the universal craving to be able to communicate our thoughts. That has led to the invention of spoken language, sign language, music, picture painting, writing, block printing, engraving, the printing press, lithography, telegraphy, telephony, photography, motion pictures, radio, television, xerography, the Internet—and the list remains woefully incomplete.

Furthermore, when we take the long view of each of these, we feel that we are looking not at invention, but rather at evolution. Each has its own huge history of individual inventions. We must also remember that none of these technologies was regarded as necessary until it had been developed to the point of being ubiquitous. Before that, each was seen, to some extent, as frivolous. For invention is created by desire before it gets around to serving need.

During those months in which we passed from the second millennium into the third, many organizations and periodicals made up lists of the most influential people of the second millennium, and one name kept surfacing—even leading—the list. It was Johann Gutenberg. Our adulation of Gutenberg says as much about the retrospective value that we place upon the printed word as it does about an individual accomplishment.

I mean to take nothing from this shadowy and somewhat erratic fifteenth-century inventor. I ask only that we view him through that wider lens, which necessarily reduces the size of components. Before we ask how (if I may put it this way) Europe finally came to invent Gutenberg, we must first determine just what it was that Gutenberg himself invented and what it was that he was trying to invent.

Johann Gutenberg was born in Mainz probably in the latter 1390s—maybe 1400.[1] His family name was Gensfliesch zur Laden. Gutenberg was the name of his wealthy father's house, derived from the words meaning "Jewish hill," since it had once been part of a large Jewish settlement. Historian John Man tells how it fell into the hands of Johann Gutenberg's great-great-grandfather after a devastating pogrom in 1282. Our inventor's name changed to Gutenberg in 1419, when his prominent family began calling itself Gensfliesch zur Laden zum Gutenberg.

Since his father was an official with the ecclesiastic mint, Gutenberg grew up knowing a great deal about the way coins were minted. One needed first to create a steel punch, then use it to imprint a gold or silver coin—techniques that clearly foreshadowed the casting of type. We can be pretty sure that Gutenberg was educated outside Mainz. And we know that he moved to Strasburg in 1434, the year after his mother died and the family estate was divided.

However, our knowledge of Gutenberg's early life remains frustratingly incomplete. Most of what we know about him is revealed only by his constant presence in courts of law. He was clearly very feisty in his dealings. Court records reveal that Gutenberg returned to Mainz in 1434 long enough to wage a quick and decisive legal battle with the city of Mainz to secure his share of his inheritance. After that, he surfaces again in a 1436 lawsuit. A mother is suing him for an alleged breach of betrothal to her daughter, an upper-class young lady named Ennelin zur Yserin Thüre. Did he lose the case? Did they settle out of court? All we know for certain is that Gutenberg so insulted one of her witnesses that, in 1437, the fellow successfully sued him for 15 guilders (or gulden).

For this, and for what follows, we need to know the value of a guilder. Man suggests that it was worth around $200 in early-twenty-first-century dollars, but elsewhere he remarks that laborer's annual salary was about 20 guilders, which suggests a value nearer $1,000. That reflects an odd difficulty in converting currency across temporal and cultural lines.

Money assumes different values as it applies to different things. I recall trying to determine the value of my dollar bills during a visit long ago in Communist Poland. It turns out that they were worth far more in bread or shoes than they were in America. But they were worth less in automobiles or pocket calculators. Similarly, fifteenth-century labor was still cheap (although it was rapidly rising in value) in comparison with the work of skilled craftsmen.[2] In any case, whether Gutenberg had to pay $3,000 or $15,000, his outburst cost him dearly.

It was just after this that Gutenberg emerged as an inventor, and once again, the story must be pieced together from a lawsuit in combination with known events surrounding it. It seems that four partners, Gutenberg, Hans Riffe, Andreas Heilmann, and Andreas Dritzehn, went into the business of selling goods to the faithful who in 1439 would be making a large pilgrimage to Aix-la-Chapelle (present-day Aachen).

And what would pilgrims need? One item had become very popular— a wide-angle convex mirror that one could hold up as one stood upon the city walls, and with which one could collect the holy rays of all the relics in the city at once. While that might sound laughable, it is a significant manifestation of a new fascination with optics in Europe. (More on that in a moment.)

Gutenberg had come upon a new process for manufacturing such mirrors, and the group had raised 1,000 guilders for it (some significant fraction of $1 million). Then several things happened. The plague reappeared, delaying the pilgrimage by a year. Dritzehn discovered that Gutenberg was hatching a second secret art for making money, along with the mirror project. Riffe exited the group. And Dritzehn fell ill and died.

All this comes to light because, in 1439, Dritzehn's family showed up in court trying to claim his portion of the investment in the project. Gutenberg got off with a 100-guilder payment to the family, but now the mystery is whether the group had turned its attention to that second "secret art" of Gutenberg's, and what that art was. The court records refer to purchases of lead and the construction of frames. All that aligns with what we have now come to know about printing with movable metal type.

Older sources are fairly confident that the group had given up on mirrors and gone into the business of printing indulgences.[3] Indulgences were, technically, extra-sacramental exemptions from punishment for sins that had been forgiven. They had become verbose documents, requiring a great deal of effort to write out by hand, and selling them had turned into a lucrative racket. If one could produce printed, church-approved, fill-in-the-blanks indulgences, labor costs would plummet, making them even more profitable than they already were.

Later historians, however, are doubtful. For one thing, we have no physical evidence that Gutenberg printed at that time, and evidence from the ensuing 1440s is slim. A fragment of text dealing with the Last Judgment, part of an astronomical calendar, and Aelius Donatus' brief Latin grammar text all appear to have been printed before 1450.[4] And all are anonymous.

The other difficulty is that the real trade in indulgences began in 1451 when Pope Nicholas authorized their large-scale sale to raise the money he needed to hire mercenaries to fight the Turks.[5] That specifically translated into Cardinal Nicholas of Cusa's order for two thousand indulgences to be created (by whatever means) in Mainz. Gutenberg was certainly printing indulgences soon after, but did he print before that?

The court case brought by Dritzehn's family also suggests that presses of some kind had existed, and that, to keep their secret safe, Heilmann had taken the precaution of having them broken down and moved. The partnership, however, broke up, and in 1442 Gutenberg moved back to Mainz, where he very clearly embarked on the project of creating printed books.

We are thus left in a dilemma. From a bibliographic standpoint, the oldest document we can actually trace to Gutenberg is an indulgence printed in 1455, the same year he began production of his Bible.[6] But our interest is in process, not product. If we mean to talk about the origins of invention, we have to go back to Strasburg in the late 1430s.

However primitive Gutenberg's first experiments (or early attempts) might have been—however far from a marketable product they still were—the long, complex process of inventing printing with movable metal type almost certainly began, in some way, with his Strasburg partnership.

Indeed, we might well be surprised that only a decade later he had advanced far enough to enter a serious, large-scale printing venture. In 1449, he was in a position to borrow 800 guilders from a Mainz businessman named Johann Fust for the project of printing an entire Bible.

The Bible was a huge effort. Each copy consisted of 322 large sheets of paper that had been folded into 1,286 printed pages. A great deal of handwork was subsequently done on each page—initial letters, fancy artistic flourishes. All that, and the book's binding, lay downstream from the printing.

The resulting Bible was almost identical to the best of the manuscript Bibles, and those had been selling in the neighborhood of 80 guilders. Gutenberg printed nearly 150 copies of his Bible, most on expensive paper imported from Italy, with some 40 or so on parchment. It cost roughly 20 guilders to print the paper ones and 50 guilders to print parchment copies. He looked to sell them at considerably less than manuscript Bibles and still gross several million in today's dollars.

What is strange about the whole endeavor is Gutenberg's almost fanatical perfectionism in carrying it out. Man draws our attention to Gutenberg's design of his pages in the classical shape of the golden section. The printed area within that page also formed a golden section; however, he split the print area into two narrow side-by side columns, each only five to seven words wide. You or I would find that annoying, but it is unlikely that many readers covered five hundred words per minute in silence, as we expect to do. Reading aloud, which can be done only at a third that rate, was far more common.

> A Golden Section is proportioned so that:
>
> $$(C - A)/A = A/C$$
>
> C
>
> Solving this algebraically gives:
>
> $$C/A = 1.6180 \ldots$$
>
> A

Gutenberg honed and summarized the style of handwritten manuscripts. He hired goldsmiths to cut almost three hundred type characters. At the time, the alphabet had only twenty-three or twenty-four letters

Fragment of a right-hand column of a Gutenberg page. Notice periods and hyphens dangling over the right-hand margin of each column, and the narrowness of the column.

(depending upon what you count as a separate letter), but he had to do much more than merely doubling that to obtain upper- and lowercase letters. For printing to look like handwriting, letters needed all kinds of ligatures, shorthand symbols, and diacritical marks.

Yet, despite his orthodox design, Gutenberg surprises us by inventing right-hand justification, something scribes had not attempted. In fact, he went further: he allowed hyphens and periods to extend beyond the justified edge. That gave his columns a nice solid look, but it also meant typesetters had to include a blank piece of type at the end of each row that did *not* end in punctuation. Later printers soon gave up on that costly nicety.

Man estimates that Gutenberg's Bibles required six compositors, twelve printers, and six presses. At first they printed forty lines per page. Then, to save money, Gutenberg went to forty-one lines by shaving his line spacing. He finally squeezed in forty-two lines. That was the rare moment when Gutenberg let cost-cutting trump perfection.

And so Gutenberg was a puzzle. He displayed what seems to be an almost intemperate perfectionism. But that was accompanied with a strong enough interest in profit that he took time off from printing his Bible to print a large batch of indulgences. For all of Gutenberg's life, however, profit appears to have been directed toward that one product, his Bible—his true north, his magnet, his grail. It, was perhaps, the sought-after oasis of creative beauty in the center of a troubled raucous marketplace of a life. Forty-eight of them survive today, thirty-six on paper and twelve on vellum, and their beauty causes us to catch our breath.

Gutenberg Bible facsimile, Houston Museum of Printing History.

As Gutenberg was finishing production of the Bible in late 1455, his debt to Fust had risen to 2,020 guilders. Gutenberg by then was probably approaching sixty years of age, and just as he was finishing his masterpiece, Fust foreclosed on him. Fust took over Gutenberg's presses, his equipment, and his fine craftsman colleague, Peter Schoeffer.

Historians have weighed Fust's action ever since. He may have committed one of the smarmiest business deals in history. But more was going on than meets the eye. The secretive Gutenberg appears to have been diverting borrowed money into a separate venture that he had no intention of sharing with Fust. He was using the money to prepare a printing of a Latin Psalter just as soon as profits from the sales of his Bible had taken care of Fust's investment.

Whatever the merits of the case, Fust did take over, and in 1457 it was he and Schoeffer who published the Psalter that Gutenberg had been working on. From then on their names, not Gutenberg's, graced the lovely books they printed.

Fust and Schoeffer's printer's device, used to identify the books they printed. (This does not appear in the Gutenberg Bible.)

Gutenberg's story makes our head swim when we try to divine the motives behind his actions. One must, I suppose, begin with the seemingly obvious fact that he was always engaged in fights over money. I know many people who seek after monetary gain, and I must say they generally make a better job of it than Gutenberg did. Gutenberg would cut down any tree, destroy any friendship, burn any bridge to keep his creative projects in motion. In that he was utterly ruthless. His were the actions not of a person interested in amassing wealth but of one who would do anything to keep his creative work alive.

Gutenberg's story leaves a question hanging in the air. Was he a visionary? On one hand, we do not have a shred of evidence that he saw where his creation would take him. Yet I would credit him with vision of a more important kind. He recognized, and acted upon, the public craving for reading matter that had been accumulating in northern Europe.

Every invention is an attempt to create a future, and so it becomes a prediction of the future. Yet we know that no future can be predicted. Thus the most visionary inventor is the person who reads the present accurately. Such an inventor builds an unknowable future upon that reading. Gutenberg did just that; he read the present with pinpoint accuracy and, having done so, hurled his vast creative abilities and impulses into precisely the right arena.

It is far beyond our present scope to trace the buildup that made efficient means for creating reading material something that had to be invented. We leave that to the Renaissance specialists. Rather, let us select a few illustrations that might give us an anecdotal sense of that buildup. A good place to begin might be the object of the delayed pilgrimage of 1439—the pilgrimage to Aix-la-Chapelle that set Gutenberg and his group into motion. Aix-la-Chapelle had been the seat of Charlemagne's empire. Because he had subsequently been sainted, it was the home of his reliquary.

Charlemagne had done a remarkable thing 658 years earlier. In AD 781 he invited an English scholar, Alcuin of York, to his court.[7] Alcuin had been in Europe the year before, and Charlemagne had recognized that he was very bright. He asked Alcuin to bring learning to the kingdom of the Franks.

Alcuin was then forty-six—a scholar/cleric who made an odd match for the worldly and boisterous Charlemagne. Yet Charlemagne knew that education would be the key to building a new post-Roman civilization in the European wilderness. According to legend, Charlemagne went at book learning like a kid with a new toy. When he met the great works of St. Augustine and St. Jerome, he asked Alcuin why he could not have a dozen or so scholars of their standing in his own court. "What!" cried Alcuin. "God himself had only those two and you want twelve!" A fine spirit of play marked their unbalanced roles of student and master, king and servant. Alcuin told Charlemagne to watch his language—to behave himself. Once he wrote about Charlemagne, "Behold our Solomon, resplendent with the diadem of wisdom. . . . Cherish his virtues, but avoid his vices."

The story has it that Charlemagne took it in good grace and bent himself to the task of learning to read, although he never quite got the hang of writing. In fact, he may never have learned even to read. That might have been a business that he left to the secretaries that surrounded him.

Here we need to recall the words of Rousseau (Chapter 1), who said that we have to "know correctly a state which no longer exists, which perhaps never existed," if we hope "to judge our present state correctly." For the results justify the legend: after developing his own education for five years (in whatever way), Charlemagne issued the first of three proclamations to all clergy. The clergy, too, were to take up letters and learning.

The clergy were an uneducated lot, but Charlemagne kept the heat on them. His proclamations were a peculiar fusion of his own authority and Alcuin's fine classic prose. They reflected the alchemy of two personalities adopting one vision.

One of the many consequences of this collaboration was a distinctive Carolingian script, developed in Charlemagne's court and named after him. This side effect of Alcuin's influence affects us today, for it was

Example of ninth-century Carolingian script, shown in the 1897 *Encyclopaedia Britannica*. Notice that our practice of starting sentences with Roman letters (or capital letters, as we now call them) had not yet begun.

made up of what have become our modern lowercase letters. The familiar capital (or uppercase) font consists of the older Roman letters, which, being less well differentiated from one another, are harder to read.

The terms *uppercase* and *lowercase* came into being after Gutenberg. Typesetters began separating their Roman letters into drawers, or cases, which they placed above the drawers containing the Carolingian letters. That way they had to reach less often into the inconvenient upper cases of Roman letters.

One of Charlemagne's proclamations was especially important. In it, he had decreed that all male children should be schooled. Of course, that was only a visionary hope twelve hundred years ago. For most of those who were touched by it, it could mean little more than learning to read parts of the catechism. It was nevertheless an important step on the road to literacy.

Alcuin retired after fourteen years. Charlemagne lived on until AD 814, after which the empire he had built fragmented. Still, his educational initiatives, while not expanded by his successors, was apparently left in place.[8] The bond between Alcuin and Charlemagne had spawned another empire that did not die. Their efforts continued to spill over into society as a whole. Charlemagne's enforcement of schooling had taken root in the monasteries. Alcuin's ideas about curriculum eventually carried into the first universities.

A poignant story about a medieval mother named Dhuoda reveals much about the climate in northern Europe during the ninth century.[9] Dhuoda was the wife of Charlemagne's grandson, Bernhard, a military and political figure in the early medieval French court. She was one of a small number of women who had taken up reading and writing. In 843, Dhuoda finished writing an instructional manual for her then sixteen-year-old son, William.

A mother seeking to raise her son by writing out a training manual seems strange indeed, but Dhuoda's circumstances were no less strange. Bernhard had put Dhuoda away in a castle, then handed William over to Charles the Bald as a hostage. He made of his son a political bargaining chip. Dhuoda, confined like Rapunzel, reached out to her exiled firstborn the only way she could—by writing her parental advice down for him to read.

Meanwhile, her husband, Bernhard, was vying for favor among Charlemagne's squabbling grandsons, of whom Charles was one. Bernard evolved into a true monster as he gained power. Cruel, lecherous, and political, he tortured and maimed his enemies, seduced the previous king's wife, and removed a second son from Dhuoda even before the child was baptized. Of course, Bernhard's enemies were no better. One by one, they had the rest of his family blinded or murdered. Charles the Bald finally beheaded Dhuoda's husband the year after she had written her book. Much good giving up his son as a hostage had done him!

And so we look at this manual—this attempt of a woman to be a mother to her estranged child. It shows a surprising grasp of the theology, philology, philosophy, and mathematics of her time. Translator James Marchand judges that it is written in fairly good, but certainly not fluent, Latin. Dhuoda speaks in a unique voice, slipping from poetry to prose and back again so deftly that one is hard pressed to find the seams.

She has a sure knowledge of the classics. She loves words, word games, arithmetic, and the mystic power of numbers. Her religious conviction is absolute, and she is fervently committed to William as his loving mother. She begins with a poem praising God and asking for William's well-being. She also spells her name out in an acrostic. Later, she calls up numbers to direct her meditations. In her thinking, the number 4 has a special perfection. It is the number of letters in the Latin word *Deus*, for God, and the first letter of *Deus* is the fourth letter of the alphabet.

Dhuoda's work is typical of scholarly medieval thinking. Yet it comes from a woman long before we expected women to be literate. While it is all rendered with the playful fluency of a fine mind chafing for somewhere to go, her playfulness evaporates at the end. She finishes the book with her own epitaph:

> *Dhuoda's body, formed of earth,*
> *Lies buried in this tomb . . .*
> *O King, forgive her sins . . .*
> *Great Hagios, unlock her chains . . .*
> *Almus, give her rest . . .*

When he was twenty-three, young William had proven to be, in the words of one old chronicle, "too much the son of [his father] in flesh and in habits." So Charles the Bald, already having beheaded the father, now had William beheaded as well. Dhuoda's manual on how to grow up in the grace of God had done scant good without her presence behind it. However, Dhuoda did leave us a glimpse into the heart and mind of one woman who escaped her veil of anonymity by means of an expanding use of the written word.

Charlemagne, of course, had been playing catch-up ball. A strong intellectual movement had followed the brushfire spread of Islam. That movement had begun after the Prophet Muhammad's death, 158 years before Charlemagne engaged Alcuin, and it was centered on Baghdad. Charlemagne had his eye upon that center and could only begin the long process of emulating it.

At first, the intellectual center of the world had tilted from Baghdad not toward northern Europe but toward Islamic Spain. By 1085, Spain had for centuries been divided into Islamic and Christian portions, and each portion in turn had separate administrative units. That year, King Alfonso VI of Spain, who had been struggling to reunite the scattered kingdoms of Christian Spain, succeeded in negotiating the surrender of Islamic Toledo as well.[10]

Christian occupancy did not last—not yet. But for a year Alfonso called himself the "Emperor of the Two Religions." His administration allowed not only Moslem and Christian to live together in peace but the significant Jewish population as well.

In the brief time before Alfonso was driven out, European scholars descended upon the library at Toledo and came away with a humbling comprehension of how far ahead of them Spain was. For four centuries Arab scholars had been the chief preservers of the ancient Greek books. Many of the classics had become distorted echoes in the Christian world, and when Europeans saw this legacy firsthand, they were taken aback by its brilliance.

The Arab and Jewish scholars of Toledo graciously led scholars from the north through their trove of literature. Now they saw much of the old literature in its original form. The works of Aristotle in particular presented them with a kit of logical tools that opened a stunning array of capabilities. They rediscovered the syllogism—that miraculous means for using two facts to generate a third. For example:

Seven is a single digit number.
All single digit numbers are less than ten.
Therefore, seven is less than ten.

Using a kind of mental legerdemain, we have suddenly generated a third fact, seemingly out of thin air.

The French scholar Pierre Abélard seized on the new logic as he turned Aristotelian dialectic loose on Holy Scripture. "By doubting we come to inquiry," he said, and "by inquiring we perceive the truth." Abélard wrote four rules for inquiry:

Use systematic doubt and question everything.
Learn the difference between rational proof and persuasion.
Be precise in use of words and expect precision from others.
Watch for error, even in Holy Scripture.

It would be five centuries before science would put these rediscovered ideas to full use. Lawyers were first to adopt it. Theologians rode the waves Abélard had unleashed, and by the mid-thirteenth century the Church began to acknowledge and incorporate Aristotle's methods. When it did, Thomas Aquinas moved in to build a theology based upon Aristotelian logic.

The demand for copies of books being discovered in such places as Toledo rose. From the twelfth century onward, Europe's voracious appetite for books put it on the road that led straight to the invention of Gutenberg. Books would remain handwritten, copied one at a time, while any number of steps toward improved arts of bookmaking were being mastered.

One large step was a sudden increase in the scope of the Church's scriptoria. The culture of the scriptorium was particularly strong from the late tenth through the thirteenth century. By the late twelfth century monastic scriptoria had evolved into book factories. In them, expert copyists—scribes—reproduced books on a large systematic scale, unlike the earlier ad hoc copying of one book at a time.[11]

We gain some sense of how these organizations worked when we follow one medieval historian, Jennifer Sheppard.[12] She found forty or so old manuscript books that were traceable to Buildwas Abbey—a monastery that once stood in western England. Although it stands no longer, it is clear that it must have had a large library for those times. Sheppard has sorted through the manuscripts that once belonged to the abbey and which are now scattered in many places. She has set out to learn where each of these books was made. The task is huge, and she has expressed the hope that she might finish it within her lifetime.

Part of her problem is determining whether Buildwas Abbey acquired all its books from outside, or if some were produced in an in-house scriptorium. So the detective work begins. It all sounds rather like Umberto Eco's book *The Name of the Rose,* or the movie based upon it. She remarks on the faint blue lines that the scribes who made these books laid out to guide their writing. As she analyzes their work, their personalities emerge. Her favorite character is one she calls the "Flyleaf Scribe." He is the one with a great loop on his *g* and a sweeping tail on his *x.*

He shows up first in a scriptorium in Cambridge. Later, she recognizes his style in the Buildwas manuscripts. He hovers over books. You see his notations on flyleaves, his hand correcting the work of young scribes. In one book she finds him steering the hands of ten novices. All that kinship and hierarchy suggests a very large operation. But was her scriptorium located at Buildwas? When she finally manages to identify books that Buildwas did buy from other scriptoria, she confirms that scribes from her scriptorium have added their own notes to those books. Thus the scriptorium, whose existence she has now solidly inferred, must have been part of Buildwas Abbey.

Sheppard warns us to be skeptical, even of her own conclusions, since hers is a tricky business. Yet there remains little room for doubt for another reason. Our confidence in her findings rises since Buildwas was a Cistercian monastery and the Cistercians are known to have been master technologists. They did wonders with waterpower and agriculture. They defined twelfth-century high technology at a time when the production of manuscript books was one of the most complex technologies. The Cistercians did a great deal of invention related to bookmaking. They introduced both alphabetical indexing and pagination, and their development of production scriptoria created such a moneymaking business that it spilled over into the lay world.

As early as the mid-twelfth century, commercial scribes began taking part in the book-copying business as the new secular universities drove an expanded interest in copying texts.[13] By the mid-thirteenth century the so-called *pecia* system had evolved. Universities rented out packets, or *pecia,* of pages of books for the students themselves, or commercial scribes, to copy.[14]

That was one more important step in the long process that took us from books costing years of labor to books that we can now pay for with an hour's earnings—even if we work at minimum wage. Students read from pieces of books, which were being copied in pieces. It was a new cottage industry that spread the written word with increasing rapidity.

With the creation of all these new books came new technologies. Throughout the twelfth and thirteenth centuries, Europeans wrote on expensive animal skins. A little terminology here: the word *parchment* is generally used to identify the skin of a sheep, goat, or calf. It is an abbreviation for *charta pergamene* (Latin for "paper of Pergamon"), since the widespread use of sheepskin as a replacement for papyrus began in Pergamon, in western Turkey, in the second century BC.[15]

The word *vellum* is often used interchangeably with *parchment,* or it can be used restrictively for the finer skin of a young (or even in utero) animal. When the scribes of Pergamon replaced papyrus with parchment, they had to replace scrolls with the kind of flat book that we read today.

Our word *volume* for a single printed book derives from the Latin *voluvere,* "to bend, to turn." From *voluvere* we get the word *revolve* as well as *volume. Volume* originally signified a book bent around a spindle. But we form our books by printing several pages upon a signature—a single piece of vellum, parchment, or paper. We then fold those signatures of pages and sew or glue them together with other signatures. The technical term for the resulting book of pages is *codex.*

The consequences of the high cost of parchment were many. One was the reuse of parchment. People would scrape off old texts and write new ones on the same sheets, which we call *palimpsests.* We have invented optical means for reading the ghost images of the scraped-off texts, and much long-lost literature has come to light. A famous case is a volume of Archimedes' work, lost until we discovered a copy made in roughly AD 1000. It had been written over as a prayer book two hundred years later, discovered at the beginning of the twentieth century, and read only very recently.[16]

Another consequence of the cost of parchment was that scribes scrimped by squeezing more words on a page. As they began producing remarkably small and crowded texts, there naturally arose attendant difficulties in reading them.

Let me suggest that we do an experiment: Let us look about at the educated citizenry in any typical public place in America. We count ten adults, and then count the number of those ten who are wearing glasses. The fraction will be large, and those without glasses are very likely to be wearing contact lenses. These optical aids to reading are now ubiquitous. I spent the summer of 1956 hunched over a drawing board and was, by July, feeling steady nausea. It turned out that I had some astigmatism and needed my first pair of glasses. They took some getting used to. Now, fifty years later, I feel undressed without them.

The increasingly literate medieval population was no less subject to vision problems than we are today. How then did they manage? Well, my first glasses, which then struck me so strongly as a technology of my brave new modern world, had already been around in earlier versions for six and a half centuries. We find the first reference to glasses in the transcript of a sermon preached in Florence in 1306: "It is not yet twenty years since there was found the art of making eyeglasses. . . . So short a time is it. . . . I have seen the man who . . . created it and . . . talked with him."[17]

The inventor's name is not given, but we do have a clean date of origin. The first glasses were made just after AD 1286. An Arab scientist had figured out how to make a spherical lens back in 1036. When his writings reached Europe in 1266, Roger Bacon asked if lenses might help old people with weak eyes. Twenty years later, this Italian inventor managed the trick.

A complication immediately arose because, in 1286, people thought that our eyes sent out rays that bounced off the things we saw, and returned to the eye. Theologians believed that sight worked a little like radar. We were told to view God's world directly, without the distortion of mirrors and glass. The outgoing and returning rays were therefore apt be bent by spectacles. Truth would be distorted.

But utility won out. By 1300 Venetian crystal workers were in the eyeglass business. Their best lenses were ground from quartz crystal; cheaper lenses were made from glass. Crystal lenses made such a lucrative trade that crystal workers weren't allowed to leave Venice and ply their trade elsewhere once they were members of the guild.

Those early glasses either pinched the nose or were held on a stick—uncomfortable either way, but still a great improvement over the failing naked eye. As handwritten book production skyrocketed, spectacles further drove demand. And in another feedback loop, the widespread use of eyeglasses made it possible to start reducing the size of type. As more words were squeezed onto each page, the amount of expensive parchment used in making a book decreased further still, and sales continued to rise.

Eyeglasses and optics were a compelling presence in the world around a young Gutenberg—witness those mirrors with which pilgrims sought to collect holy rays of light. Thirty-nine years after Gutenberg printed his first Bible, Sebastian Brant wrote a book titled *Ship of Fools (Das Narrenschiff)*—an illustrated diagnosis of human folly in its many forms.

23549   Frameless Eyeglasses, double or periscopic convex or c o n c a v e lenses, good quality................. .50
23550 F r a m e l e s s Eyeglasses, same as 23549, extrafine finish, fitted with patent adjustable nose guard.................... .75

Detail of an image in S. Brant, *Das Narren-schiff* (Basel: Johann Bergmann von Olpe, 1494), from a Latin version published in 1497 (Special Collections, University of Houston Libraries).

A typical pair of glasses offered for sale to ordinary people in the 1895 Montgomery-Ward catalog. Take notice of the prices.

This was now a world that had largely moved to the new medium of print, and Brandt shows us a foolish scholar, surrounded by too many books. The manic fellow uses a feather duster to whip through pages faster than he can digest them. He looks like an owl in his oversized spectacles. (So, I suppose, do we, armed with our own eyeglasses, flipping through more words than we can digest, seven centuries later.)

By now a development that had arisen long before Brant and long before Gutenberg was crying out to be adopted into the burgeoning book trade. It was *paper*. One might think that common sense would have dictated a switch from the use of expensive parchment to far less costly paper. Paper had been around almost as long as parchment, but its adoption in Europe was surprisingly late. Perhaps the paper-related word *ream* offers a clue as to why it moved so slowly.

We might well reflect, when we open a new ream of shiny white paper, upon the Arabic term *rizmah*.[18] It means "bale" or "bundle." The Spanish made *rizmah* into *resma,* and the French made *reyme* of it. It finally became the English word *ream,* a bundle of twenty quires (or a total of 500 sheets) of paper. The word trail for *ream* almost perfectly marks the trail of paper as it finally began moving from the East to the West.

The Chinese invented paper in 49 BC and began using it as a writing material in AD 105.[19] By the seventh century, its use had spread eastward to Japan and westward as far as Samarkand. That was just after Islam had begun its own outward spread from the Mideast. Paper and Islam converged when Arab forces reached Samarkand, in southern Asia, just north of present-day Afghanistan. The Arabs then encountered this new material, which had been in use there for about two generations. Under their rule, Samarkand became a papermaking center, along with Arab papermaking centers in Baghdad and Damascus.

All paper is made of plant fibers of one sort or another, and Samarkand paper was made of mulberry fiber. Two strong recollections linger with me after my visit to Samarkand, many years ago. They are the beautiful mosques and the warm air, heavy with the delicate perfume of the ubiquitous mulberry.

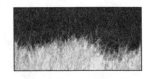

A 1/3-inch length of bond paper, torn against its grain. The fibers were scattered in a slurry, which was drained through a screen, leaving the paper sheet behind. In this case, the lay of the fibers favored the vertical direction, giving the paper its grain.

The intellectual center of the older Hellenistic world had been the city of Alexandria, located at the mouth of the papyrus-rich Nile Delta. To make papyrus, one hammers reeds that have been laid out in a crisscross pattern. The fibers are much longer than those in paper. They do not adhere by entangling, as do paper fibers, but rather are cemented by the reed's sap.

Papyrus was a fine material, but the papyrus reed was too localized to supply the rest of the world. Therefore, after Pergamon became a parchment-based intellectual center, parchment spread across Europe as the preferred writing material.

But intellectual ascendancy was passing to Baghdad in the eighth century, just as Arab scholars had discovered paper. So that ascendancy came to rest upon this new writing medium. Historian Jonathan Bloom emphasizes the importance of that fact.[20] Before we had cheap and abundant paper, arithmetic required erasing and shifting numbers—operations that could be done on slate but not on paper. In AD 952, Arab mathematician al-Uqlidisi used Indian algorithms to create once-through methods, which, since they involved no erasing, could now be done with ink on paper. To make full use of paper, we needed these same methods we use today for multiplication and long division.

The use of paper continued its slow westward migration. Cairo was manufacturing it by the tenth century, Tunisia and Islamic Spain by the eleventh. Paper did not cross the Pyrenees into Europe; rather, it entered by way of Islamic Sicily and was being made in Italy by 1268.

Paper finished its long looping trip from China through Samarkand, into the Holy Land, across North Africa, and up into Central Europe, finally reaching Moscow in 1578. That is why it seems so ironic that Samarkand was a part of the Soviet Union when I was privileged to see it. Samarkand— great nexus in the glacial migration of paper, way station of the old Silk Road, ancient, sunlit, and sweetly touched with the aroma of mulberry.

Both Hebrew and Islamic scripture had traveled on parchment, and both religions had been reluctant to put scripture on anything so modest as paper, despite its strength and durability. The flow of paper into Europe was further slowed by Christians, and their resistance was fueled by yet another factor: they viewed papermaking as an infidel technology. Central Europe did not take up the use of paper until the fourteenth century, and England only at the end of the fifteenth—well after Gutenberg.

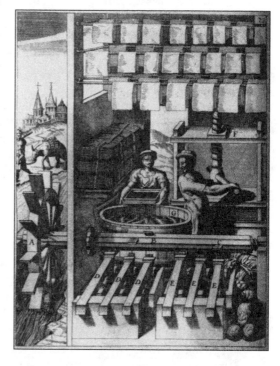

European papermaking in the seventeenth century. An undershot water wheel on the left powers a mechanism that pounds balls of rags into pulp. A man with a square screen drains a slurry of rag fibers in water to create a sheet of paper. The man next to him is pressing the sheets out flat and hanging them on rods to dry.

The capacity of the printing press could never have been fully tapped without the Chinese invention of paper, and by AD 1400 paper was available in all of central Europe. That much was ready for Gutenberg. So too was another Chinese technology for speeding bookmaking, block printing.

In the broadest sense, printing is very old. Stamps, brands, royal seals, and even imprints of fossils on limestone are, after all, forms of block printing. One of the earliest forms of Chinese printing was done by lay-ing paper upon images or characters carved into a block of stone or wood, then rolling ink over the paper. That gave a negative image, much like our present-day stone rubbing.

By the eighth century the Chinese had inverted this art of rubbing to create conventional block printing—the kind achieved by inking a raised image and pressing it against the paper. They used this tech-nique on a far grander scale than anyone before them, printing not just pages but whole scrolls. Before AD 1000, Chinese Buddhists had printed complete books of doctrine. That took 130,000 blocks of wood and twelve years to finish.

The fifteenth century found medieval scribes in the West using wood block printing to replace the work of artists in making illuminated let-ters, and the few stiff illustrations, in their books. Some even tried to do what the Chinese Buddhists had done, block-printing pages with all the words carved into single blocks. They combined such pages (cruder than the Chinese ones) into simple children's books. Others used block print-ing to make decks of playing cards with identical backs.[21] (Here, too, the Chinese had been first. They had printed playing cards as early as the ninth century.)

We know in retrospect that movable type would have to be the means for providing widespread reading material. But even as we name Gutenberg the canonical inventor of that technology, the Chinese trump us once more. In AD 1045, a printer named Pi-Sheng did almost what Gutenberg would do 410 years later.[22] He shaped individual characters on the ends of small square clay rods and aligned them, face up, in a shallow tray lined with warm wax. He laid a board across the array and pressed it down until each character was at exactly the same level. When the wax cooled he used this array to print pages.

The hardest part of Pi-Sheng's process was the huge number of char-acters that had to be created and which his typesetters had to pick up and set. The Chinese did not have a streamlined alphabet of twenty-four or so characters. Still, Pi-Sheng's printing went into use. The Chi-nese used it only sporadically, but the technology still spread. By the time Gutenberg printed his Bible, the Eastern use of printing with mov-able type had reached Korea and Japan. Some Moslem scribes were also aware of it, but it did not suit their love of hand calligraphy.

In Europe, Gutenberg was quite possibly one of several people work-ing on printing. Others could conceivably have printed one or more of the anonymous fragments that appeared before 1455. Such speculations might tempt us to sell Gutenberg short, but that would be a mistake. He ultimately created durable metal letters that fit together with a jeweler's precision to yield a truly beautiful printed page. After Gutenberg, the art of printing acquired a capacity that would permit it to bring learning to the masses. After him, an age-old technology was now poised to turn the world upon its ear.

The impact of print was probably greater even than we imagine it to have been. Between 1455 and 1501 somewhere between 8 million and 24 million volumes were printed, depending upon which expert's estimate we accept. Those volumes represented upward of forty thousand titles. That massive infusion of books was about to give rise to massive social upheaval by any reckoning.

No new technology moves that fast in a world not primed to receive it. The medieval world was primed. Europe had cried out for accessible written material since the high Middle Ages. Now, with its remarkable compact alphabet, Europe—unlike China, with its enormous number of individual characters—assimilated printed books with breathtaking speed.

Naturally, the result was social disruption. We call it a Renaissance, and surely, despite all the resulting turmoil, it was a rebirth. As all those books redirected history, they did so, in part, by redirecting invention. People began discovering much more on a sheet of paper than anyone had meant to put there. Books started to say new things to people. They began to alter thinking in ways that no one could have predicted.

Radically new ways of seeing the world arose out of the new printed books. In particular, they unleashed an array of new sciences upon the unready world. That is where we want to go next. A new external reality was about to be created, and, with it, new ways in which invention could occur.

O how sweet life was
when you and I sat—quiet
amidst all those books.

—Alcuin[23]

10

From Gutenberg to a
Newly Literate World

Gestation to Cradle to Maturation

We gain a new respect for the elusiveness of invention if we ask what people were saying about their new printed books by 1501. By that date, the technology of book printing had left its cradle and was on its way to becoming a mature technology. Seventeenth-century European bibliophiles used the date 1501 to mark the end of the period that ran from 1455 through 1500. They adopted the word *incunabula* to identify books printed during that period, and their designation has stuck. That choice was a convenient way to divide the calendar, but it was not based upon scrutiny of the innate sequence of invention.

The Romans originally derived the word *incunabula* from their word *cunae,* "cradle," to mean, among other things, "source" or "origin." Europeans turned *incunabula* into a noun intended for plural use, as in "These incunabula are all from France." Then they had to invent the singular form *incunabulum* to mean just one book. Today, book people often get around all the altered Latin by anglicizing the word—by simply using the word *incunables* to describe books from this period.

The word *incunabula* thus evokes an image of many early printed books occupying a cradle of printing, where we arbitrarily pick 1500 as the last of those cradle years. Yet we would surely expect that a period yielding fifteen million or so printed volumes had reached beyond mere infancy. And, when we look closely at the period from 1455 to 1501, we see the use of printing undergoing major transitions toward the middle of the incunabula years—perhaps around 1476. The year 1455 is also problematic since, although it marks the first commercial printed books, Gutenberg was actually on his way much earlier to inventing printing with movable metal type.

I should like to suggest three stages in the evolution of any technology, giving them the names I suggest in this chapter's title: gestation, cradle, and maturation. The *gestation* is a fairly long run-up period before the invention takes an identifiable form. It is a period during which we seriously begin working to create some capability that we strongly desire, without knowing what form that capability will take.

The *cradle* follows the first appearance of the invention in a reasonably functional form. It represents a time during which inventors hone the rudimentary engine of their ingenuity and seek to determine what it is that they've really invented. During the cradle years, the invention is altered in ways that make it more fully serve us. (Notice that by applying the word *cradle* to books, we put ourselves at liberty to select a range of dates different from those of the incunabula.)

Maturation is the period after which the form has been roughly established, and in which we and the technology begin the process of adapting to each other. The maturation is only the coming of age of a technology, not its full lifetime. Let us call maturation complete when the technology has finally reached a form that from then on is only inflected and no longer dramatically revised.

In Chapter 3 we speak of how the ancient dream of flying gave way to systematic study and invention after the Montgolfier brothers first flew humans in hot-air balloons. The resulting gestation of flight lasted until the Wright brothers put all the pieces together in 1903 and produced a rudimentary machine that could reproducibly achieve powered flight.

Between then and 1934, we sifted through many experimental forms of airplane. We came close with the Fokker and Ford Trimotors, but the required elements for fully viable commercial use did not converge until we had the Douglas DC-2 and the Boeing 247. The cradle of the airplane's evolution had been a thirty-one-year period of steady experimentation. After that, our airplane culture would come to maturity as commercial transports progressed from the DC-3 in 1935 to the first successful jet airliners. The DeHavilland Comet (the DH-106D), the Boeing 707, and the Douglas DC-8 were all in use by 1958.[1]

Steam engines reveal a similar history. One might begin the gestation with Della Porta's 1606 proposal to use steam to pump water (Chapter 4). After Della Porta, many people began trying to invent means for using steam power. The cradle began with Newcomen's engine in 1712, and it came to completion when Smeaton and Watt took up steam engine development in the 1760s.

Watt's invention of the separate condenser in 1764 makes a good ending for the cradle of the steam engine. The steam engine's period of maturation ran from then until around 1804, when Evans and Trevithick created high-pressure engines and began using them in transportation systems.

Although we do not deal in detail with the history of the computer here, we can briefly parse its evolution in our present terms as well.[2] That is especially worth doing, since the computer brings with it echoes of the evolution of the printed book that are too strong to ignore.

Like the history of flight, an interest in mechanical computation, and mechanical means for accomplishing it, can be traced to antiquity. Best-known of these is the abacus, which was so effective that, in many places, it remains in use today. However, a sudden increase of interest in finding means for mechanizing calculation accompanied the seventeenth-century scientific revolution.

Many of the great mathematicians of that era worked on creating viable calculating machines. John Napier laid the foundation for the first workable device when he conceived of logarithms. Although he also experimented with mechanically aided calculation, it was English mathematician William Oughtred who, in 1621, saw how to mechanize the use of logarithms. He invented the slide rule, illustrated below.

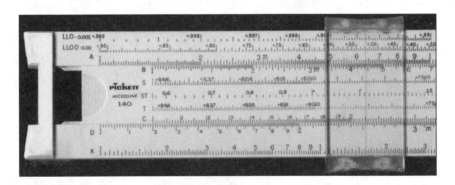

An unprogrammed calculation: 252 divided by 21 equals 12.

While work on various calculators continued sporadically for the next two centuries, the idea of programming computations—of creating a machine to execute a string of instructions—was another matter. It appears not to have crossed anyone's mind until Charles Babbage proposed it in 1834. That ab initio event is where the gestation of the modern computer began.

We reached the cradle of programmable computation in 1930, when Vannevar Bush finally built the first successful computer.[3] His huge analog/mechanical machine had, by World War II, grown until it utilized 2,000 vacuum tubes and 150 electric motors in its operation. Bush called it the "differential analyzer."

Four elements, all radically new, had to be made a part of computers before they could leave their cradle: binary digital computation, internally stored user programs, transistors to replace electromechanical elements and radio-type vacuum tubes, and integrated circuits.[4] These were all in place by 1958.[5]

A computer economy matured over the next twenty-five years. Computers became an integral part of record keeping in business and commerce; programmable computation became ubiquitous in academia and research laboratories. But other threads also emerged. The large mainframe began giving way to the personal computer. The Internet also evolved during this period until, in 1983, it could be made available to home users of personal computers. That was also roughly when lay computer users began buying shrink-wrapped software packages in stores. Computers had matured by 1983.

That left us, the general public, facing a very steep learning curve, but climb it we now would. I had my first personal computer in 1982 and was using the Internet by 1992. I jumped through hoops along the way; but, after the mid-'80s, it became increasingly clear that the payoff for making such an effort would be immense.

To parse the evolution of the technology of the printed book in these terms, we begin with its gestation, which we describe in Chapter 9. It started with the rise of systematic book production. When? I suggest a date of roughly 1230, which is the point at which Humphreys describes a secular scribal trade spilling out of the monasteries onto the streets.[6] Gestation ended when Gutenberg (and his coworkers?) developed a system of printing with movable metal type. That appears to have taken place around 1438, well ahead of the Gutenberg Bible.

It is important to emphasize that all these dates are rough and subject to debate. I make no claim to precision, and recognize that any might be tinkered with this way or that. However, each of these dates should be robust enough in the face of challenge to illustrate my intent in using the words *gestation*, *cradle* and *maturation*. They are listed in Table 10.1 for printed books, steam engines, airplanes, and computers.

Table 10.1
Gestation, Cradle, and Maturation of Technologies

Technology	Gestation Dates / Length (years)	Cradle Dates / Length (years)	Maturation Dates / Length (years)
Printed book	1230–1438 / 208	1438–1476 / 38	1476–1525 / 49
Steam engine	1606–1712 / 106	1712–1764 / 52	1764–1804 / 40
Airplane	1783–1903 / 120	1903–1933 / 30	1933–1958 / 25
Computer	1834–1930 / 96	1930–1958 / 28	1958–1983 / 25

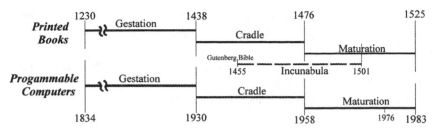

Stages in the evolution of printed books, and of programmable computers

Two of these proposed dates remain to be explained. They are 1476 and 1525. To see why I suggest them, let us return to our opening question: What were people saying about their new printed books early in the period of maturation—in 1501? And let us to respond to the question in terms of an analogy between the book and the computer.

We might expect to find people reeling from the impact of 15 million new printed books, but they were not, even though the social order was on the way to being turned inside out. In 1517, Martin Luther would post his Ninety-five Theses on the Wittenberg church door, and what he posted would be a printed document that could be widely circulated.

The print-driven scientific revolution of the sixteenth and seventeenth centuries also lay largely in the future. Upheaval was, for now, rather like a tsunami still out in open water. It would stay invisible until it breached its shoreline. By 1501, 15 or so million books had been flung into a world where scholars had recently expected to travel miles for the privilege of visiting a library of twenty hand-written volumes. Now, even as they were reading books in their own homes, they had only hints of the future being shaped by all those books.

The situation causes me to look back at 1976, forty-six years after Vannevar Bush's development of a programmable computer. What was I thinking then? Fifteen years earlier, I'd finished an engineering Ph.D. without ever being in the presence of a digital computer. At that time, the early vacuum-tube mainframes had been around, but only the cognoscenti got near them. By 1976, that had changed. I was teaching at the University of Kentucky, and numerical computations had become a normal part of my work.

Yet it was in 1976 that a directive came down mandating that all campus computation would henceforth be done on the campus mainframe computer. The messy proliferation of small stand-alone computers was to cease. The maturation of the programmable computer was about to be completed by affordable PCs, and we were oblivious to the fact that the earth was shifting upon its axis. The new books had been pouring into the world of 1501 in much the same way that mainframe computation had penetrated into ours. And the response they received was not unlike ours in 1976.

Still, the late fifteenth century was not totally silent on the matter of the new printed books. Printers themselves were first to comment on the changes being wrought by print. They speak to us from their explicits.[7] An explicit was a personal note that a scribe was allowed to write at the end of a manuscript when it was completed. It comes from the Latin *explicitus,* which means "having been completely unrolled" (as a scrolled papyrus book). The complementary meaning of *explicitus* was "fully revealed."

Manuscript explicits simmered with personal intensity. Scribes used them to curse anyone who might steal the book or in any way defile it. They would tell readers to wash their hands and not let their thumbs smear the ink. They complained about their cramped backs and stiff necks. They also implored readers to treat their work with respect. I especially like an explicit in which a twelfth-century scribe looks back upon his months of labor and utters a twofold prayer that it will be treated well and that it will serve some reader's mind:

> This book, O Christ, in praise of thee,
> Lies finished for all to see.
> Good Benedictine, to spare my health,
> Put back this book upon its shelf;
> And you will give me recompense
> If you deem it worthy of your intelligence.[8]

In 1501, it was customary to include the information that you and I expect to find on a title page, in a colophon at the end of the book.[9] But many early printers had taken to expanding upon the bare production information of a colophon and adding more personal notes—thus turning colophons into printed explicits. In them, they wrote about how many books they could now make, and how accurately they could now make them. Of course that was largely advertising, not intended as social commentary. A few German writers, fed up with the cultural superiority emanating out of Italy, used their expanded colophon/explicits to boast about German printers. In 1498, Sebastian Brant wrote this interesting commentary, not in an explicit, but in a poem:

> In our time, thanks to the talent and industry of those from the Rhine, books have emerged in lavish numbers. A book that once would have belonged only to the rich—nay, to a king—can now be seen under a modest roof. . . . There is nothing nowadays that our children . . . fail to know.[10]

Notice that Brant primarily addresses the *availability* of books. Does he see how all those books are pulling the threads out of the medieval social fabric? He does say "there is nothing nowadays that our children . . . fail to know." But he does so in the same gee-whiz voice that we so often applied to computers and the Internet throughout the latter twentieth century.

In fairness to Brant in 1498, or to myself in 1976, it has never been possible to diagnose accurately any revolution going on around us. And our understanding of the print revolution is further obscured by the fact that we try to read it from records that appear in print. The situation is reminiscent of what philosophers call the "error of self-reference"—the logical trap of trying to make statements about any system from within that system.

We have already seen how hard it is to trace Gutenberg's and Dritzehn's primary invention of printing with movable type (or, for that matter, Newcomen's and Calley's invention of the atmospheric engine) from written records. It is particularly difficult in the case of Gutenberg and his early collaborators, since the printed book had yet to emerge as a means for recording history.

From all we know of the cradle phase of other technological evolutions, we look for intense development accompanied by a great deal of modification and experimentation. The cradle is a time during which inventors ring changes upon the original invention, varying its form and seeking out those improvements that will prove viable.

Yet the print media of the late fifteenth and early sixteenth century provide little in the way of descriptions of current printing technology, and only a few include images of presses. When images of printing presses do appear, they generally look much the same in every case, and they are attributed to only Gutenberg and/ or Fust and Schoeffer. Historian James Moran suggests that this reflects the lack of records of the intervening evolution.[11] We have a fairly clear picture of printing technology at the turn of the sixteenth century, but little sense of just how it got there.

We do deduce from the extraordinary quality of Gutenberg's Bible that he had solved many of the thorny problems involved in

An excellent replica of a latefifteenth-century "Gutenberg press" made by Steven Pratt, Pratt Wagon Works (Museum of Printing History, Houston, Texas).

doing high-quality printing. How many of his solutions match those of a Gutenberg press as it is generally represented today, we cannot be sure.

A few of those mechanical problems that had to be solved by Gutenberg as well as those who followed his were:

- Creating better means for founding the type than coarse sand casting. Gutenberg brought to bear upon this problem his special knowledge as a silversmith who had grown up with an intimate knowledge of techniques used in a coin mint. His method resulted in a vast production of what might well be viewed as the first fully interchangeable parts. It was a huge technological accomplishment— perhaps the most significant one he made.
- Inventing means for positioning the tray of type very precisely in a solid container (called a *coffin*).
- Mounting and aligning the type. This involved precision shimming, once type had been cast to extremely close tolerances.
- Accurately placing the coffin under the press before applying pressure. That came to be done by sliding it in on a "carriage" that rode on rails.
- Briefly but intensely pressing the paper against the inked type. This meant significantly altering existing screw press mechanisms. It also meant that means had to be created to keep the mechanism from transmitting any twisting torque to the paper.
- Developing inks that adhered, rather than flowing as ink must do in a pen.
- Finding means for applying just the right amount of ink to the type.

The tip of the screw in a "Gutenberg" press is a blunt toe that exerts pressure without transmitting torque. Below are the carriage rails (Museum of Printing History, Houston, TX).

Such a list could soon be stretched to book length, especially if we included all the sub-problems that might be listed under each. While Gutenberg had clearly found his way around these problems, we have no evidence that his solutions all took exactly the form of the canonical "Gutenberg press" as early as 1455.

Take the single matter of positioning the paper over the solidly aligned tray of type. By 1501 that was done with the help of a tympan and frisket. They are the two upward-tilted rectangles that appear at the left of the photo of a complete press (page 163). The lower rectangle, the tympan, holds the paper. The upper one, the frisket, is a mask that prevents any ink outside the type area from touching the paper. These items are not shown in the earliest drawings of presses.

Variations on these functions were undoubtedly still evolving rapidly in the near wake of Gutenberg. We also have evidence that many wooden parts of presses were being replaced with metal castings. The cradle years were not idle ones.

For about thirty years after Gutenberg improvements in the technology of book making were focused on expanding the capacity for production. Printing presses were moneymaking machines whose purposes remained much the same as Gutenberg had originally conceived them—mass-producing the classical and religious literature in the style of the old manuscripts.

It took a second generation of printers to begin making radical changes in the purposes of printing—to begin using books to accomplish a large variety of things that did not at first occur to anyone. The changes that took printing out of its cradle, and marked the beginning of its maturation, included the widespread creation of literature in local (non-Latin) languages, the creation of a new secular literature, and the ultimate reorganization of knowledge that would have to occur once it became reproducible.

One particularly dramatic change was the rise of a new concept of scientific illustration. Scribes had generally been unable to handle any detailed or carefully made illustration. The same image of an old man might be stamped on a finished page to represent Moses and then reused to represent Aristotle. Images had been included not to represent but to evoke. Besides, no copyist could have handled a detailed watercolor of a mandrake root or an accurate rendering of a human skeleton.

Simple woodblock image of a strict tutor and four pupils, used repetitively by William Caxton in several books during the late fifteenth century. Caxton's pioneering work in print did not enter the new realm of illustration.

This simple iconography at first continued in printed books exactly as it had in manuscript books. Printers (and authors) finally tumbled to the fact that they could make one carefully drawn woodblock image of a real thing and use it in a print run of 250 books. (More on that in Chapter 11.)

Changes such as these redirected the use of print in the middle of the incunabula period and began the maturation of a technology that was now fairly well honed. Yet most printers, scholars, and readers failed to see where printing would take us, just as in 1976 we failed to see where our programmable computers would take us. And, as is so often the case, when radical visionaries arose, they arose from outside the mainstream. Let us meet just one of those people—the person who introduced printing to England and redirected its purpose in an English-speaking world.

He was William Caxton, and we might start his story in 1474, with the remarkable explicit he added as he finished his three-volume translation of Raoul Lefèvre's histories of Troy, *The Recuyell of the Historyes of Troye*. Caxton's explicit sounded like something any scribe might have written—a simple cry of fatigue: "My penne is worn, myne hande wery ... myne eyen dimed."

Caxton, however, had *done* something about his wearied hand and dimming eye. When we read the full context of his lament we're in for a surprise. We discover that he had taken up arms against the limitations not only of his tired and aging body but of a pen-and-ink technology that would tax anyone's body. Here, in this section of his explicit (with the spelling updated), you and I find ourselves joined with Caxton in a moment that altered history:

> Thus end I this book, which I have translated after my author as nigh as God has given me cunning; to whom be given the laud and praising. And forasmuch as in the writing of the same my pen is worn, my hand weary and not steadfast, my eyes dimmed with overmuch looking on the white paper, and my courage not so prone and ready to labor as it has been, and that age creeps on me daily and enfeebles all the body; and also because I have promised to diverse gentlemen and to my friends to address to them as hastily as I might this said book: Therefore I have practiced and learned, at my great charge and dispense, to ordain this said book in print, after the manner and form as you may here see. And it is not written with pen and ink as other books be; to the end that every man may have them at once. For all the books of this story, named the "Recueil [compendium] of the Histories of Troy," thus imprinted as you here see, were begun in one day, and also finished in one day.[12]

Caxton had indeed "practysed & lerned ... to ordeyne this said book in prynte," for it is in the medium of print that we read this astonishing passage. However, his remark about finishing in one day needs qualification: Printers of the incunabula typically did a full run of one signature of pages, printed on one side, in a day's time. In that sense it is quite true that the run of pages "as ye here see were begone in oon day and also fynyshid in oon day." Once I had learned his story, I no longer begrudged him that small bit of hyperbole.

Caxton had been born in England in 1416. In 1442, he had gone to Bruges, across the English Channel in the duchy of Burgundy, as apprentice to a merchant.[13] He did well in that great European cultural center. By the time Gutenberg printed his great Bible, Caxton was wealthy and had begun collecting manuscript books. He also rose to a position of political clout in Bruges, eventually serving for a time as governor of the English merchants in Bruges.

By 1469, he had entered the service of Margaret, duchess of Burgundy, right after her marriage to Charles the Bold, duke of Burgundy. Margaret was King Edward IV's sister, and the marriage had been one of pure political convenience—a marriage in name only. Margaret was a noted scholar and arbiter of good taste in literature. She was only twenty-three and very intelligent. There appears to have been a natural affinity between her and the avuncular scholar Caxton, thirty years her senior.

At this time Caxton was working on his English translation of Raoul Lefèvre's three-volume history of Troy. Margaret liked it very much and pressed him to complete the series. In fact, Margaret kept asking Caxton for copies of the first two volumes, which had proven very popular in her court. Caxton had to write each copy out by hand.

Caxton finally solved his problem by learning the new art of printing. He appears to have spent 1471 and 1472 in Cologne, learning about printing and, perhaps, commissioning a type that he liked. As a printer, however, it seems pretty clear that he learned some and invented the rest.

As nearly as we can date Caxton's history of Troy, it came out early in 1474. The quality of the printing seems crude alongside Gutenberg's craftsmanship. Yet Caxton saw his work in a larger frame of reference, one that allowed him to reshape the very purpose of literature as he printed.

His typefaces, for example, might first strike one as being coarse and showy alongside the austere classicism originally created by European printers. However, his type provided better differentiation among the letters, and that made it much easier to read than Gutenberg's. It offered the same kind of advantage that lowercase Carolingian letters provided over uppercase Roman letters.

After Caxton had printed his third book, he left Burgundy and, around 1476, set up England's first printing press at Westminster Abbey. At the age of sixty, Caxton set off on a brief but extraordinarily fruitful fifteen-year career as printer. He worked with furious intensity

Quod cũ audiſſet dauid: deſcendit in

But among a thouſand cheſe the

Comparison of type from a Gutenberg Bible, above; and Caxton's *Dictes and Sayenges of the Phylosophers* (1477), below. Adapted from the 1970 *Encyclopaedia Britannica*.

Caxton's printer's device. This elegant shape has become a motif for contemporary jewelry.

and was preparing his one hundredth title for the press when he died in 1491.

At Westminster, Caxton became a major agent in changing the purpose of printing. The early printers had given their patrons what they had expected: beautiful and costly copies of Latin manuscripts. Now Caxton gave them what they both wanted and needed, a diverse, practical literature, both secular and sacred, largely in their own tongue. He provided his readers with English history, *The Canterbury Tales*, sayings of philosophers, a book on manners, a book on chess, a French-English dictionary, and romances. What he had begun with his Trojan history he finished with a book on the use of arms and the formalities of chivalry.

Europe still regarded England as a relative cultural backwater in the late fifteenth century. The fact that Caxton was operating outside the circle of European high culture served him well. From his outside-the-box vantage point, he did far more than just take up printing. He brought populist literature to the people. Without a Caxton there could not have been a Shakespeare.

While Caxton provides one (certainly not the only) point of departure between the cradle and the maturation of printed books, we might ask of him, as we asked of Gutenberg, "Was he a visionary?" He surely took a huge step toward radicalizing the print revolution when he took books to the people. A cynic might shrug and say that Gutenberg was just trying to make money and Caxton merely suffered from writer's cramp. But that misses the driving creative intellect that carried each to the true center of the problem he faced.

Technology is like that. We do not understand the synergy between ourselves and our machines until we are able to view that synergy in hindsight. And it takes a very peculiar kind of vision for an inventor to interpret the problem of function in such a way as to bring about a result that lies beyond his own, or anyone else's, predictions.

Before we see how the maturation of printing played out, let us leapfrog ahead, for a moment, to a point beyond 1525—to a time when the democratization of reading was far more complete, as the result of Caxton's and other's efforts.

Carlo Ginzburg tells the tragicomical tale of Menocchio in a book with the tantalizing title, *The Cheese and the Worms*.[14] In it he introduces us to the Italian miller Menocchio. Menocchio provides a startling picture of the sixteenth-century world as it had been recently transformed by the ubiquity of books that nonscholars could read.

The miller, born in 1532, used Menocchio only as a nickname; his real name was Domenico Scandella. He was a friendly, loquacious fellow, always shooting off his mouth. He was well known in his small town of Montereale. We get some idea of his turbulent relationship with Montereale when we find that he was exiled from it for two years after he got into a brawl at the age of thirty-two, and that at the age of forty-nine he served for a year as its mayor.

Menocchio had developed considerable fluency as a reader, although he was in no real sense educated. He simply sensed that all those new books held wonderful secrets. And so he read and he talked. He spent his precious money on precious books. He swapped books with literate friends in other towns. The words made only a kind of patchwork sense to him, but the bits and pieces of learning obsessed him. He collared friends on the street to harangue them about such matters as the Trinity and the Virgin Birth. He had particularly interesting ideas about our cosmic origins.

Ginzburg is able to tell us a great deal about Menocchio and about just what he had read, since he had access to court records (more on that in a moment). Menocchio's readings were all done in his native Italian. Indeed, he boldly complained against the use of Latin. He called it a betrayal of the poor, because it left them helpless in any legal proceedings. That, of course, was an implicit criticism of the Church.

So it can be no surprise that the Italian Inquisition finally hauled him into court. "Keep your mouth shut," his family told him. Menocchio tried, but it was no use. He faced a panel of real scholars who were ready to listen to what he had to say. He faced a notary who would write down every word. It was a dream fulfilled. Menocchio talked—and talked. His theology violated all orthodoxy and contradicted itself. It was earthy and filled with rich metaphors.

Ginzburg derives the title of his book from Menocchio's remarkable cosmology. Menocchio's beliefs about our origins were based upon a theology of putrefaction. The heavens, he said, had formed when the vast primordial chaos curdled into planets and stars, as cheese curdles out of cream.

Angels came into being in this ferment the way worms seem to appear spontaneously in rotting cheese. Indeed, all living beings appeared from the chaos of the firmament in this way—angels, humans, and the greatest being of them all, God himself. The details tended to vary during Menocchio's trial. Sometimes God preceded the chaos. Sometimes he was manifested from within it. Academic consistency was not his strong suit. Yet the chaos, the curdling, and the worms remained constants within his formulation.

Hubble photograph of galaxies curdling out of the deep field some 13
billion light-years from the earth.

Now, as telescopes reach the far fringe of the universe, all that sounds
a bit less silly. For now we see pictures of stars curdling out of the chaos
of some thirteen billion years ago, almost exactly the way Menocchio
imagined it.

As Menocchio talks, we hear random echoes from all the great for-
bidden books of his age—the Vulgate Bible, the *Decameron*, the Koran.
His anger with the Church for controlling all that knowledge kept spilling out. He was suspected of being a "Lutheran." He had certainly read
about that new sect in the north, and he might well have been influenced by their concept of personal revelation. But he made it known
that he, like the court, regarded Lutheranism as evil.

The court found Menocchio guilty of heresy and threw him in prison
for life. Three years later, a sick and thoroughly chastened Menocchio
convinced authorities he had changed. They released him, and he went
back to work. But those beautiful new printed books were still there,
and he could no more ignore them than we can, in the long run, avoid
television or cell phones. He was soon talking again.

This time the trial was shorter. This time they subjected Menocchio,
now sixty-seven, to torture on the *strappado* in an attempt to get him to
name people with whom he had talked. Menocchio slyly gave them only
one name—the lord of Montereale, whom the court was not about to
mess with. Since the court blandly reports that torture had been applied
with moderation, we can probably assume that his baffling stubborn-
ness temporarily saved him from a worse fate.

Ginzburg believes that Menocchio was doing more than just protect-
ing his friends. His condemnation of the Lutherans probably had more
to do with asserting his passionate hope that people would realize that
his cosmology and his theology were his own. And, even under torture,
he continued to assert that he had borrowed none of his cosmology from
others. Rather than credit others with his thoughts, Menocchio, it appears, would have preferred to die, and that, of course, is what did happen. In 1599, the Inquisition (spurred by Rome's rising interest in the
case) finally ordered him burned at the stake.

He had, in fact, been no part of any reformation but his own. He went to his death fearful and protesting, as anyone would. However, he remained proud of the thoughts he had thought to the end. *Of course* he had baffled the court. Through the medium of free and unconstrained reading, he had become a creature of a world incomprehensible to those who had not entered it along with him.

The crowning historical irony is that, today, we read the old Inquisition court records *in a printed book.* Now Menocchio speaks to scholars once more, and he does so from the pages of a printed book. Books have allowed him to outlive his inquisitors. Today, he really does charm scholars, and he finally takes his own place in the very medium that changed history at the same time it was spelling his doom.

So with Menocchio in mind, let us look about us once more at over a billion computers that have been thrown into the world during a scant two decades. Like Sebastian Brant, we tell one another, "Gee whiz, look at all the information our children can now access." The real changes that the computer is bringing about—changes in the way we see reality—remain invisible.

We hardly yet have an adult generation that has known the personal computer from birth. At this writing, you and I still see the computer against the background of the not-computer. We typed before we word-processed. We learned the algorithms of arithmetic before we used hand calculators. We memorized facts, algorithms, and spellings.

All of us see the personal computer against the backdrop of a world without it. What we cannot see at all is how a mind will work when it has never known anything else. What did they say about books in 1501? In the end, whatever was said was irrelevant because it was—ipso facto—useless commentary. For everyone looking at the new books in 1501, the future was as hopelessly unpredictable as it remains today.

11

Inventing Means for Illustrating Reality

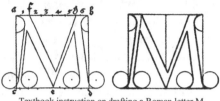

Textbook instruction on drafting a Roman letter M
Left: Dürer, 1525.   Right: Svensen, 1941.

Maturation of the printed book, as we have described it, took about fifty years. One thread of its development, more than any other, lay at the heart of its coming of age: the evolution of visual means for representing information.

One cannot help but be reminded of *Alice in Wonderland*. Young Alice sits by her sister next to the rabbit hole, and "having nothing to do: once or twice she had peeped into the book her sister was reading, but it had no pictures or conversations in it, 'and what is the use of a book,' thought Alice, 'without pictures or conversation?'"[1] Alice was perfectly describing the books of the cradle years of print.

In Chapter 10 we found William Caxton doing something about the absence of Alice's "conversations" in the books of the incunabula. He created an English-language literature that people could read for pleasure. Pictures would enter right on the heels of "conversations," and the two drawings above offer a subtle hint as to how things went after Caxton (and others like him) had begun broadening the scope of the new medium.

Each of these two letter *M*'s is a textbook display of how to draft the letter in its proper classical Roman form. Albrecht Dürer gave us the first in 1525, the year that I propose as a rough boundary for the maturation of printed books.[2] The other has nothing to do with early printing; it appeared 416 years after Dürer's, in a nineteenth-century engineering drafting textbook by Carl Svensen.[3] Yet it is practically identical to Dürer's.

In the mid-twentieth century, Svensen presented the design of twenty-six elegant Roman letters for use as titles on engineering drawings. He acknowledged Dürer for having first provided these designs, although Dürer had rendered only twenty-four of them. The *J* and the *U* had not been part of the old Roman alphabet, and the final content of the

Western alphabet had yet to settle down even in Dürer's time. Svensen added his own versions of those letters, drawn to be consistent with Dürer's guidelines.

In our word-processing world, we pay increasing attention to the appearance of our documents. We select among fonts once available only to professional printers. We have become mini-publishers, deciding among Courier, Bookman, Century, and other fonts. One of the most ubiquitous of these is Times New Roman, although calling it Roman is a stretch. It includes both the old Roman capital letters and versions of the subsequent Carolingian lowercase letters, done in a compatible style.

If we compare Times New Roman uppercase letters with Dürer's, we see that they are remarkably similar. Actually, all the classic serif fonts in use today echo Dürer, for he set the standard around which they continue to orbit. But the kinship with the popular Times Roman font is especially close.

AGKMY
AGKMY

Above: The Roman capital letters as Dürer drew them. *Below*: The Roman capital letters in the Times New Roman font. Dürer used slightly thinner lines, and this letter *M* (one of three he designed) has vertical legs, making it a bit narrow.

Dürer's Roman letters appear in his treatise on the use of geometry in representative art. Dürer dedicates the book to a friend and patron, the humanist scholar Wilibald Pirckheimer. In his dedicatory preface, Dürer laments the way earnest young artists produce junk because they have failed to learn the geometry of perspective and proportion. His lengthy section "Of The Just Shaping of Letters" begins thus: "Now, since architects, painters & others at times are wont to set an inscription on lofty walls, it will make for the merit of the work that they form the letters correctly."[4]

Dürer sets down general precepts for forming letters. Each is to occupy a square. Thin lines are one-thirtieth of the dimension of the square, and the heavy ones one-tenth. Decorative serifs on the letters are formed about arcs, and their diameters are also specified as fractions of the square. He then goes on to give specific instructions for each letter. For each he describes a geometrical construction to be carried out with a compass and a straightedge.

His instructions are not original. During the preceding half century, a number of Italian artists had been suggesting various specifications of this kind for forming the Roman letters.[5] Dürer's specs were the last word, not the first, on forming Roman letters. It is by virtue of their extraordinary grace that they are the ones that endure today.

But Dürer had done something much more than design a typeface. His letters encapsulate his understanding of the combined esthetics and mathematics of composition. They are beautiful because he had an astonishing understanding of balance and proportion. His far more famous etchings and paintings incorporated exactly those same principles.

When I first studied drafting and descriptive geometry, though I did not realize it then, I was studying Dürer. Not only were his letters among our exercises, but I also learned to lay out sheet metal patterns and design duct intersections in accordance with his instructions. Perhaps I should not be surprised at that. When I read Dürer's foreword, I discover that it had been his intent to teach me—right along with the inexperienced young artists of his own day. He says just that, making direct reference to the engineers of his day:

> Wherefore I hope that no wise man will defame this laborious task of mine, since with good intent & and in behoof of all who love the Liberal Arts have I undertaken it: nor for painters alone, but for goldsmiths too, & for sculptors, and stonecutters, and woodcarvers, and for all, in short, who use compass and rule, and measuring line—that it may serve to their utility.[6]

Even after World War II, engineers still reached, now and then, for Dürer's Roman letters when they wanted to put a classy finish on their drawings, but the world was growing increasingly utilitarian. Dürer's letters kept appearing in our textbooks, but his instructions for laying them out were more and more often omitted. They finally vanished entirely from drafting instruction, and today we merely download a good approximation.

Dürer's 1525 treatise was the great *summa theologica* of his thinking about geometry and art. But he seems also to have reached a sort of closure when he returned, late in life, to the art of typefaces. By then, the use of illustration and imagery in print had reached an astonishing level of sophistication, and Dürer had emerged as a great pioneer of these new visual arts. His return to the formation of letters, just three years before his death in 1528, seems almost to have been a recognition of—a tip of the hat to—the way he had bridled the force of print and wed it to his art.

We noted, in Chapter 10, that representative illustration (at least as we would recognize illustrations today) did not initially appear in early printed books, even though manuscript books had occasionally been alive with pictures. In medieval bestiaries, for example, we find remarkably lifelike elephants, beavers, and hyenas. But they were one-of-a-kind drawings, not reproducible prints. And they appeared alongside equally wonderful phoenixes and unicorns.[7] The desire for reliable images of external reality was clearly present, but what Dürer accomplished in the medium of print began only in the years after Caxton had begun printing at Westminster Abbey.

In 1484, six years before Caxton's death, Johannes Veldener, a printer in the Flemish city of Louvain, published an illustrated herbal.[8] Herbals, or books on how to identify specific flora, were the first subject of the new fully representational block prints.

rriiij

𝕳 brotanū Auewnde

es ſeet inden ierſten Dzoeg in
dē twedē ﬁ.teghēt wytual
lē des ſhers dats alopitia ef
ſcozfeſhept of ſchellingſhe des
ſoets dats tinea es auewnde

Detail of a Veldener page, after Hellinga.

Buglossa ochsenzungen

Buglossa siue lingua bonis quod idem est quia assi
milatur in figura sua lingue bonis. 7 virtus eius pri
ma est que calefacit et humectat. Secūda virtus est
que ofert tussientibo et asperitate pulmonis. Isto
modo fiat potus. R. succi buglosse. lb. 1. ysopi. mel,
lisse ani. m. ç. radicis lilii. radi. yreos. enule campane
liquiricie ani. ʒ. ç. ficuum siccap numero. vij. buliant
singula cū aq. lb. ij. ad ofsumptōʒ medietatis colatū

A recognizable oxtongue plant in Petri's herbal. Woodcut with watercolor.

And here we see some of the interweaving of creative change that was affecting the new medium of print, for Veldener was the same printer who developed the third version of Caxton's distinctive typeface. Veldener now turned around and used that same font in his herbal.[9] Veldener's herbal was one of the earliest books to negotiate the shift from printed icons to authentic representational art.

In Johann Petri's herbal, published two years after Veldener's, we can see how simple line drawings were giving way to more descriptive prints. His woodcuts now left room for a watercolorist to hand-color the images in each individual copy of the book.[10]

A generation later, in the early sixteenth century, herbals had reached a level of representation that could easily stand up to twentieth-century standards of scientific illustration. By then (the early sixteenth century) people expected illustrations to tell them how things really looked. And the new descriptive science of botany came into being, propelled by a rising public interest in such visual material.

Representation of a geranium in a leaf from a Bohemian herbal, ca. 1516.

Despite the quality of the Bohemian herbal leaf shown above, it is an image that could still be rendered with no understanding of perspective. Once we had control of perspective, the range of experience that we could communicate would make yet another great leap forward.

Questions about the substance of invention arise again in the context of perspective. The ancient artist who understood that a figure of the same size must appear larger in the foreground and smaller at a distance had already begun to grasp something of the meaning of perspective. The Florentine architect Brunelleschi wrote some of the geometrical rules of vanishing points by the turn of the fifteenth century.

The Arnolfini Wedding by Jan van Eyck, 1434, and a detail of the spherical mirror reflection.

Artists such as Masaccio and Uccello soon began putting his ideas to use. But even without that mathematical understanding of perspective, a northern contemporary of theirs, Netherlands artist Jan van Eyck, must also be credited as a pioneer of perspective. In the years just before Gutenberg, when the new Italian ideas had yet to reach central Europe, he had begun producing drawings that captured complex three-dimensionality almost perfectly.

Van Eyck's 1434 painting *The Arnolfini Wedding* includes an astonishing show-off flourish of perspective. On the wall behind the young couple, he paints a spherical mirror. The near-perfect perspective in the room continues into the mirror, where, now spherically distorted, it reveals their backs as they greet us—we viewers become two fifteenth-century friends of the couple, just entering the door that the couple faces.

Two centuries later, artists may have used such mechanical aids as the camera obscura in their work. But van Eyck accomplished what he did using neither mathematics nor machines. He simply had an uncanny ability to reproduce what he saw. People were looking for new ways in which they might record reality, and he tapped into that impetus.

That ability, of course, was spotty, and it would remain so for a while. Some artists got it; most did not. For fine examples of an imperfect understanding of two-point perspective there is no better place to look than in Hartmann Schedel's *Chronicle of the World,* published in Nuremberg in 1493.[11]

Schedel's quixotic attempt to tell the history of the world utilized 1,821 images, including would-be pictures of all the great cities. And since nothing could demand more intricate perspective than a city, these images of cities convey a very clear sense of just how far woodblock artists had generally come (and failed to come) in their ability to represent reality.

Best known of the artists who contributed to Schedel's *Chronicle* was one Michael Wolgemut. In 1486, seven years before the book was published, Wolgemut took on an apprentice. He was the fifteen-year-old Albrecht Dürer, who was himself a Nuremberger. Dürer worked under Wolgemut for four years. We do not know for certain whether he was involved with any of these city images, although stylistic analysis suggests that he was not.[12] Still, he was certainly present while the woodcuts, with their imperfect perspective, were being made.

The city of Trier, as represented by Schedel.

After his apprenticeship with Wolgemut, Dürer spent time in Florence, where he began to shed the more Gothic overtones in his work. He lived in Basel for a while. There we are fairly certain that he worked on the woodcuts for Sebastian Brant's *Ship of Fools*. He was quite probably the lead man in a crew of four or so woodcut artists.[13] (Although he probably did not do the remarkable drawing of the fool with the owl-like glasses, shown in Chapter 9, it was the work of a member of his team.)

Soon after his time in Basel, Dürer took up a new medium—one that would prove very important not only in his own later work but also in shaping the directions that print would subsequently take us. It was copperplate engraving. Copperplate was less in evidence than woodcuts were during the maturation of print. Yet it came into use at the same time that woodcuts were finding their new popularity. It represented the second of the three essential forms of printed images, the third being lithography.

Woodcuts are a form of block printing, in which every part of the surface that is to be inked, and which is intended to touch paper, is raised. Copperplate engraving is a form of intaglio. The word *intaglio* comes from the Latin *intagliare*, "to cut into," and refers to the process in which an image is cut into a plate. Ink is spread on the plate, filling in the cuts, and the excess is wiped away. The incisions then hold just enough ink to leave their mark on the paper.

Lithography, however, would not be invented until the nineteenth century. It is a process in which a surface (originally of stone) is treated chemically. Ink adheres only where it has a chemical affinity. One inks the plate, then wipes off the excess ink. That leaves just enough ink in the areas where the plate was treated to make an image.

  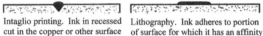

Blockprinting. Ink on a raised portion of the type, or the carved illustration   Intaglio printing. Ink in recessed cut in the copper or other surface   Lithography. Ink adheres to portion of surface for which it has an affinity

The metal type used by all the early printers executed a form of block printing. When early printers began including woodcuts in their trays of type, they were actually block printing with both images and type at the same time. Just before Gutenberg, however, a number of European artisans (goldsmiths in particular) had begun developing the process of printing with engraved copper plates. Soon after Gutenberg began working on printing, an anonymous Flemish artisan was making decks of cards using the intaglio process.

Several European engravers had produced some very fine engravings by 1496, the year Dürer made his first engraving. But Dürer was about to leap far ahead of them. His father had been a goldsmith, and he was intimate with that art/technology. It was natural that Dürer should take up engraving, and of all his works of art, his engravings are the best known.

Engraving allowed artists to create highly detailed images far more easily than they could in wood. It also demanded that printing presses impose greater pressure than block printing required. When we wonder whether an image in an old book was block printed or engraved, one of the "tells" that indicates a picture is an engraving is an indentation of the paper surrounding the image. The high pressure needed to create a clean image creates that bas-relief border.

Armed with this new ability, Dürer returned to Italy for two years, from 1505 to 1507. This time he focused especially upon the Italian knowledge of perspective.[14] When he came back and began recasting perspective in the language of solid geometry, his art took on a hypnotic three-dimensionality.

We see his full mastery of the new medium of copperplate in his late engravings. Our eye roams those pictures from detail to detail, through layers of symbolism, then back to the whole. Look, for example, at Dürer's image of St. Jerome in his study. The depth of field is astonishing. As our eye takes us into the picture, it is hard not to feel that we are physically

St. Jerome in His Study, 1514.

present within the spaces. Our interest rides from element to element the same way it might follow a fine storyteller.

In 1519, we find an image of St. Anthony huddled in the foreground of a Dürer engraving. A city reminiscent of the city of Trier in the drawing from Schedel's *Chronicle* rises up the hill behind Anthony. This time, however, Dürer, almost off-handedly, renders the buildings of the city in the background in the complex three-point perspective that had been far beyond the grasp of his teacher Wolgemut.

Only thirty-five years after Veldener's crude representation of an herb, art and geometry had become bedfellows. Another six years and

St. Anthony, 1519.

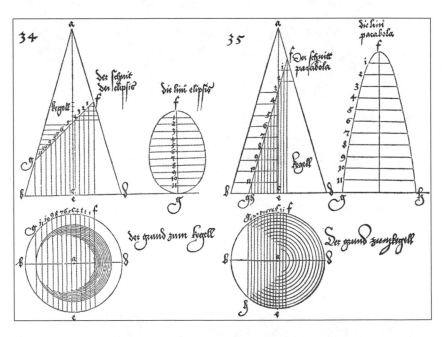

Dürer's graphical derivations of the forms of ellipses and parabolas from a cone. This is a typical construction in the subject of descriptive geometry. Two centuries later, the kindred topic of analytical geometry (which treated the same matters using algebra and calculus in place of geometrical construction) would emerge as separate from descriptive geometry.

Dürer wrote his treatise on all that he had learned about the fusion of art with Euclidian geometry. That 1525 book on constructions with compasses and rulers included the Roman letters with which we began this chapter. It also included many means for replicating reality with mechanical devices.

Like Dürer's Roman letters, one might see this as old wine in a new bottle, since the Italians had already done so much with it. However, his treatise perfected a language of exposition that set the foundations for what would be one of the first branches of applied mathematics. It was the subject that we engineers called *descriptive geometry* as we used it in the twentieth century, in the same form Dürer had given it to us.

Under the driving impetus of print and evolving means for printing images, the new science of botany had arisen during the late fifteenth century. Now we had the cornerstone of modern engineering drafting as well.

erhaps we can better understand the full meaning of the transition that had taken place by our approximate date of 1525, if we return for a moment to an early printed book and look at the initial letter on a typical page. This block-printed letter *P* is from a 1482 printed book on Euclidean geometry—a book that sends a complex message.[15] The text itself has only crude block-printed line diagrams to illustrate Euclid's constructions. Although this is now a printed book, the geometrical drawings remain sufficiently simple that a scribe copyist previously would have been able to include them by hand in a manuscript book.

In the older manuscript tradition, initial letters were treated much differently. They were far more complex than the simple geometric figures, and a scribe simply would have left a blank space in his finished handwritten page. He would have passed the page on to an artist with the skill to draw them.

That, at least, is how things were done until the very late days of manuscript bookmaking, when block-printed elements began turning up as means for saving labor and cutting costs. As the production of printed books took off, printers continued to include very fancy initial letters; but now, far more often than not, they block-printed them into the typeset texts of their books.

These ornate letters had a larger purpose than mere decoration. Like so much medieval illustration, they invited readers to engage the eye of the mind—led them to keep their inner vision keenly focused. Sebastian Brant's fool with the feather duster and the too-large glasses sounded a significant warning in 1494. He reminded users of the new printed books what could happen (indeed, what really has happened) when we read without introspection.

Not many of us today have looked at a manuscript page and fallen into near-hypnosis, mentally tracing the vines and tendrils woven about an elaborate initial letter, and more is the pity that we have not. For we cannot see the world accurately if we do so only through a lens, regardless of whether that lens is biological or mineral. Our mind always tells us things that are beyond the reach of our eyes.

Poets, artists, composers, mathematicians, and engineering designers all know that. At their best, they fuse the two visions. We need only to go back and look at that engraving of St. Jerome in his study to know how well Dürer had wed his inner and mystical vision with his exterior and analytical perception of objective reality.

But we had now invented illustration, and the internal eye would come under a slow and relentless assault over the next three centuries. The compensation was that, as we now began accurately reporting our observations in many copies of books, a whole new array of observational sciences would rise up. Botany and descriptive geometry had been only the beginning.

Next would be anatomy, and here the inner eye would continue to serve and expand upon what the observer saw. During the fifteenth century, the new representational arts had fed a rising interest in descriptive anatomy. Leonardo da Vinci finally took anatomy to a new level.

Around 1506, just when Dürer was honing the geometry of representation and the use of engraving, Leonardo began seriously dissecting large living creatures. He cut up some thirty human cadavers and a far greater number of large animals, his favorite being the ox. He left upward of eight hundred anatomical drawings.[16] Leonardo clearly recognized the importance of images. In a world where the classical anatomies described the body in words, he knew the picture absolutely had to be the medium for recording what we learned.

Leonardo said, "With what words [can you] describe the whole arrangement . . . the more detail you write concerning it, the more you will confuse the mind of the hearer."[17] So he cut and he drew. He had no illusions about the work: "You may perhaps be deterred by natural repugnance, and if this does not prevent you, you might be deterred by fear of passing the night hours in the company of these corpses, quartered and flayed and horrible to behold."[18]

Leonardo created a new pictorial vocabulary for the body's inner contents, which at first seem to be undrawable. He invented cutaway sections; he made his own X-rays by laying semitransparent tissue over underlying organs and bone. Leonardo the artist resurrected the dead and gave them new life, even as Leonardo the dissector cut them into pieces.

He was first to identify the sinus cavities, first to identify the heart as a muscle, and first to see that it was four-chambered. He showed us that every skeletal muscle had another muscle working in opposition to it. He used castings to determine the shape of internal cavities. And he said, "If you wish to demonstrate in words ... do not meddle with things appertaining to the eyes by making them enter through the ears, for you will be far surpassed by the painter."[19]

Yet Leonardo failed utterly in one terribly important way: he was fundamentally secretive and never did embrace the new medium of print. His anatomy was an artistic tour de force that had limited influence during his life and had almost no role in redirecting medicine. That would have to be accomplished twenty-four years after his death (and fifteen after Dürer's) by the Belgian doctor Andreas Vesalius.[20]

Vesalius studied medicine in Paris and then moved on to Italy. He had trained in a medical tradition that was still medieval—largely based upon the principles set down late in the second century AD by the Roman doctor Galen of Pergamon.[21]

Galen had studied anatomy in Corinth and then Alexandria. He honed his anatomical understanding by doctoring the gladiators of Pergamon. From them, he learned a great deal about the flow of blood and other fluids, abdominal organs, muscles, and nerves. But, aside from that early surgical experience and his study of human skeletal remains, Galen learned most of his anatomy by working on apes and monkeys—a fact that seriously colored many of his findings.

Still, his vast writings sustained him as a fixed medical landmark for over a thousand years. Although he had used the question-asking methods of Aristotelian empiricism, he consciously and aggressively set up a static and unquestioning authority in his own name.

In a Galen-based anatomy class of Vesalius's time, a professor on an elevated platform would typically read from Galen's texts. On a table below, a barber-surgeon cut pieces from a cadaver and a demonstrator showed them around. The relation between the hacking below and the Latin text above often became comically tenuous.

Vesalius was bright and impatient. As he watched his third dissection, he grew so frustrated that he finally snatched the knife from the barber and went at the cadaver himself. After that, he continued to dissect cadavers and exposed a vast number of elements of Galenic anatomy that differed from those in the human body. Unlike Leonardo, Vesalius used only human cadavers—no oxen. And unlike Leonardo, he took his work to press. By the age of twenty-eight, he was the leading anatomist of his age when he wrote the first modern text on the subject: *On the Workings of the Human Body in Seven Books* (*De Humani Corporis Fabrica Libri Septem*). He finished it in 1543.

Vesalius, who understood where art had been headed, worked with artists (probably students of Titian) in Venice. Artists and anatomist together created a new expository vocabulary of medical illustration. Their sumptuous detailed woodcuts were better than Leonardo's sketches (and much more accurate), as they revealed human bodies in successive stages of autopsy.

Vesalius's images of flayed bodies first strike us as macabre, but then we look more closely. When we see his image of an opened-up human skull, our eyes are drawn to the somber visage below. This is not just an open skull, for it has a human face; and that face is disquietingly reflective of what we hope we are and what we know we shall become.

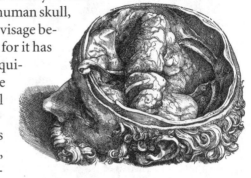

We realize that, in Vesalius's time (a century after Gutenberg), observation was merely supplementing the eye of the mind. Later, the mood would favor replacing it entirely with cool objective detachment. But not yet.

From A. Vesalius, *De Humani Corporis Fabrica Libri Septem* (Basel: Johannes Oporinus, 1543).

Vesalius provides yet another glimpse into the relation between print and invention in the text of a short book titled *The Epitome*—a short summary version of his *De Humani Corporis Fabrica*. The petulant tone of his dedication to Prince Philip suggests that he might be angry at having to write such a summary. An abridged version of two very long rambling sentences gives the flavor of it:

> [We know] how much is lost in all sciences by the use of compendiums. Though indeed they seem to provide a . . . systematic approach to the perfect and complete knowledge of . . . that which is set down with more space and prolixity . . . , nevertheless compendiums do signal injury [to] literature; for given [their] use alone, we read scarcely anything else . . . all the way to the end.[22]

He adds, "I pass over in silence those pestilent doctors who never even stood by at a dissection," whereupon we immediately hear echoes, both of our own friends who rail at the superficiality of the Internet and of Sebastian Brant's fool turning pages with the feather duster. Although Vesalius provides a brilliant text, he nevertheless despairs, from the outset, of telling the story through the medium of words. Leonardo had been right. Pictures would now have to carry the narrative; pictures would become ascendant.

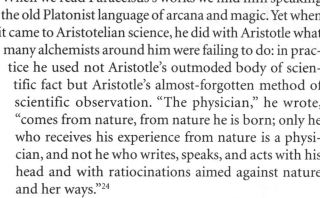

An event that took place sixteen years before Vesalius published his anatomy helps us to see that book in perspective. In 1527, at the University of Basel, in Switzerland, the bombastic alchemist Paracelsus threw Galen's medical textbook into a bonfire.

Paracelsus's given name was Philippus Aureolus Theophrastus Bombastus von Hohenheim. Born in Switzerland in 1493, he learned medicine and metallurgy. He also became a wanderer—a scholar without a classroom teaching the alchemy that underlay medicine and metallurgy alike.[23] He was fractious and difficult, going from town to town, offending people until they sent him packing to the next town.

When we read Paracelsus's works we find him speaking in the old Platonist language of arcana and magic. Yet when it came to Aristotelian science, he did with Aristotle what many alchemists around him were failing to do: in practice he used not Aristotle's outmoded body of scientific fact but Aristotle's almost-forgotten method of scientific observation. "The physician," he wrote, "comes from nature, from nature he is born; only he who receives his experience from nature is a physician, and not he who writes, speaks, and acts with his head and with ratiocinations aimed against nature and her ways."[24]

A Vesalius skeleton.

When he was thirty-four he reached Basel. There the printer Froben lay ill and no doctor had been able to help him. Froben heard about Paracelsus and, having nothing to lose, summoned him. Under Paracelsus's care, Froben did indeed rally. That event forms a convoluted but important link in our story, since Froben set books for the great humanist scholar Erasmus, and no one understood the potential power of the new medium of print better than Erasmus, who has been called the first "spin doctor."

Erasmus, as well as Froben, engaged Paracelsus as his personal physician, and Paracelsus seemed to have found a place in Basel. But then his bombast got the better of him. He lit his bonfire and threw Galen's writings into it. He might have survived the faculty's rage had Froben not died after all. When he did, Paracelsus's protection died with him. He spent his last fourteen years wandering, and governments banned his books.

Yet, in the end, Paracelsus conquered history. Under his madness had lain a huge sanity. He had told doctors to be more careful in dosing people with mercury. He had begun reshaping Earth, Air, Fire, and Water into something more consistent with corporeal chemical elements.

He had said that doctors had to analyze symptoms and monitor medicine more carefully.

Paracelsus was the sort of alchemist we read about—a magus. Yet he transmuted the language of magic and alchemy as he spoke it. He became the greatest alchemist of all just *because* he was not what he appeared to be. He was, in fact, a true creature of the Renaissance. When others spoke of white magic and black magic, Paracelsus said, in essence, "No! There is only *one* magic. Good or evil vests in the practitioner, not the art." With that, he laid responsibility upon the magician, and *magic* became another word for a property of nature.

Paracelsus demanded observation. Reading antiquated texts could not possibly be enough. He told doctors that the evidence of their eyes, digested in their heads, would teach them what ancient texts could not. And since he wrote in German instead of Latin, people read his works as they turned to the new medium of print. The printers saved him from himself after all.

Now back to Basel for a moment, for Froben was not the only printer who helped Paracelsus when he needed help. Paracelsus picked up an important young disciple in Basel in the person of the young printer Johannes Oporinus, and they made a very odd couple. Paracelsus almost killed Oporinus by experimenting on him. Oporinus honored Paracelsus's abilities but added that he was irreverent, a glutton, and a drunk. All the while, Paracelsus was writing patronizingly about "faithful Oporinus" and incurring such a maelstrom of chaos as to finally drive Oporinus away.

Two years after Paracelsus died, Vesalius went to Oporinus to have his own anatomy text printed. Indeed, Vesalius went to great effort to have his heavy woodblock illustrations transported across the Alps from Venice to Oporinus's shop in Basel.[25] (That, by the way, underscores the reason that Vesalius still wrote in Latin. This book was meant to speak to scholars, and it would have to cross linguistic boundaries to do so.)

In any case, here is Oporinus straddling the great gulf that separated Paracelsus and Vesalius.

Oporinus's remarkably fanciful printer's device—a man riding a dolphin and playing a rebec.

Paracelsus not only spoke in the old language of alchemy but also remained an iatrochemist—a medieval pharmacologist. He never would have countenanced dissection and, in fact, believed that "the dead and decaying body had nothing to say to the living body."[26] Vesalius, on the other hand, exemplified the new observational science that would dominate the coming seventeenth century. Paracelsus and Vesalius, at once so different and yet so much a piece of the same revolution—I can only wonder how Oporinus, living in the eye of the hurricane, looked back upon these two giants at the end of his life.

Vesalius had found the right person to print his book. Throughout the early sixteenth century, Oporinus had built a track record of printing dangerously revolutionary material.[27] And when he printed *De Humani Corporis Fabrica,* he gave us a work that not only spelled a medical revolution but also was an exquisite work of art.

Still, change of this magnitude never occurs overnight. Vesalius had forged the link between medicine and anatomy that had been missing, but he was still a Galenic practitioner. The ancient doctors would continue to guide medicine for at least another century.[28] Anatomy would establish its way in medical practice through a back door.

Let us, therefore, return to the barber-surgeons who did the actual cutting in a college of medicine. The word for surgery at the time was *chirurgery*—an adaptation from the ancient Greek word χειρ, "hand." *Chirurgery* meant a practice involving use of the hands. The very language in Vesalius's day set surgeons apart from academic medicine by tagging then as manual laborers. Before we could invent modern medicine, we would have to bring those lowly surgeons in under the tent of medicine.

Let us therefore meet Vesalius's contemporary, the French surgeon Ambroise Paré, born in 1510.[29] Paré apprenticed in the brutal trade of a barber-surgeon—first in the provinces, then in Paris. At the age of twenty-six he became a military surgeon. He was pitched into a seemingly endless series of nasty wars, all of which were being waged with a new weapon—firearms. Shot and shrapnel had come of age during the fifteenth century, and they were now making war more terrible than ever.

Paré found his eureka moment soon after as he served during the French attack on Turin. The enemy had repulsed the French with muskets, among other weapons, and Paré set about to treat the wounded. The Galenic physicians for whom surgeons worked were doctors with clean hands. They simply assumed gunpowder to be poisonous, and prescribed that gunshot wounds should be cauterized with a boiling mixture of oil and treacle. Thus soldiers who were already horribly damaged were scalded with boiling oil.

The siege of Turin found a young Paré facing too much carnage. His cauterizing oil ran out. As a stopgap, he tried a cold mix of egg yolks, oil of roses, and turpentine. That accident of war set up two test groups. Later in the evening, he could not sleep for worrying over his experiment. He got up to look in on the wounded. He found that

> those to whom I had applyed my digestive medicine, to feele little paine, and their wounds without inflammation or tumour, having rested reasonable well in the night: the others to whom was used the sayd burning oyle, I found them feverish, with great paine and tumour about the edges of their wounds.[30]

Paré suddenly knew that the center of his job as a barber-surgeon was more than preserving cannon fodder. It was to ease suffering, and a patient in pain could teach him what the mind alone could not. From then on, this gifted and compassionate observer created a new and humane concept of surgery. He wrote simply and directly about what he saw. He also invented all kinds of new surgical instruments and techniques. Like Paracelsus, he wrote in the vernacular—in French, not Latin. And his books soon found their way into English translations.

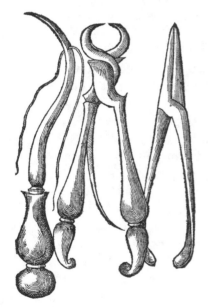

Instruments for suturing with wire, from A. Paré, *Three and Fifty Instruments of Chirurgery.*

Paré also used rich illustrations of his external vision. He was a pure Aristotelian observer, no longer wearing any trappings of academic alchemy or of Neoplatonism.

Religious conviction fed Paré's medical conversion. He invented a phrase that's been worn down to a pious cliché. When he repeatedly said, "I dressed the wound, but God healed him," I read that mantra as the wisdom of seeking out his own ignorance and dealing with the facts as they presented themselves to him.

He became the lifelong enemy of firearms, writing in his *Apologie and Treatise* of "this hellish Engine tempered by malice and guidance of man. . . . Canons, Bastards, Musquits, feild peices, . . . Basilisques, Sackers, Falcons, Falcononets . . ." He concludes, "Wherefore we all of us rightfully curse the author of so pernicious an Engine."[31]

Late in his life Paré wrote a book on birth defects called *On Monsters and Marvels.*[32] Here we again read the tension between alchemy and modern science. The old medieval rhetoric that was absent in his surgical writings seems to be creeping back in. He begins by listing the causes of birth defects: the glory of God; the wrath of God; too much seed; too little seed; corrupt seed; mingling of seed; indecent posture by the expectant

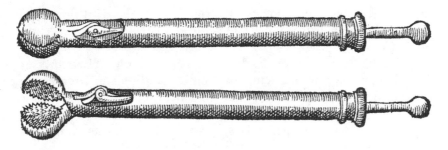

A bullet puller, from A. Paré, *Three and Fifty Instruments of Chirurgery.*

mother; a narrow womb; a blow to the mother; demons; devils; and finally, the mother's imagination.

That done, he recounts a terrible gallery of case histories, including Siamese twins and people born with missing or extra body parts. He includes the stuff of fables, such as a man's head on a horse's body, but when he does, he always identifies that as ancient hearsay. He does assert, without evidence, which cause begat the malformation. A two-headed child was the result of too much seed, and so on. Yet he takes pains to debunk many monstrous birth defects as tricks used by beggars.

And so, as we trace Paré's odd rhetoric we catch a glimpse of the real texture of shifting thinking. Indeed, we are reminded of the language of Paracelsus. We realize that while he writes in the voice of his times, he also leads himself, and us, away from the view of the world that style reflects. He weaves the methods of modern observation into a work that starts out looking like a scrapbook of contemporary folklore. This is the way the world shifts under our feet; this is how change occurs while we may glimpse it out of the corner of one eye, or we may miss it altogether.

So many of the old books fool us in just this way. It is hard for us, with our twenty-first-century eyes and rhetoric, to see the huge changes rumbling beneath a leftover prose style. We incline to separate Vesalius from Paracelsus. We want to put Vesalius in the camp of modern science because he broke cleanly with alchemy. In the case of Paré, we must be very careful not to classify him by the language to which he reverts.

The book-driven Renaissance was necessarily only the beginning. What began after the printed book matured would come to fruition in the seventeenth century. In his 1620 book, *Novum Organum,* not only would Francis Bacon declare war upon the old alchemy, but he would also describe the modern experimental method, the form of science widely recognized even today.[33]

The inner eye had revealed as much as it could without new tools. When Gutenberg embarked upon his new commercial enterprise of printing, he unwittingly provided the right tool at the right time. Two centuries later, the external eye had evolved a language with which to report what it had seen.

But in the time of Dürer, Vesalius, and Paré, the impact of print was still pretty embryonic. Academic medicine would remain largely centered upon Aristotle and Galen, while the subculture of barber-surgeons was quietly morphing into something that would begin to fit our contemporary concept of surgery. The surgeons and anatomists would ultimately reclaim possession of medicine, but not quite yet.

Along with them, other new visual sciences would emerge: zoology, descriptive ethnography, cartography, and geology. Engineering would be presaged by large and elegant books describing all the machines that one might build, along with many—such as perpetual motion machines—that none would ever build.

Naturally, the pendulum would swing too far and it would have to return. Science would have to rediscover something of the old alchemical vision. The great thinkers of the Renaissance all knew that. Leonardo, Dürer, Paracelsus, Vesalius, and Paré all made it very clear that what came in through the eyes had to be fully processed in the head and the heart. In the end, we need to be reminded that anything less yields a distorted understanding.

Science evolved in some remarkable ways while the Aristotelian eye, the outer eye, our corporeal eyes, were ascendant. Then it began to founder. Much science began to look like large grab bags of undifferentiated fact. Finally, at the end of the eighteenth century a new force arose to reclaim what we were beginning to lose. The Romantic poets told us that we create nature by dreaming nature.

That sounds like crazy talk at first. Then we catch their meaning. The time had come to reclaim our inner vision. In his poem "Tintern Abbey," William Wordsworth helps us to understand what it means to create reality within our minds, and then to cast its shadow upon the corporeal realities that surround us:

> . . . that blessed mood,
> In which the burthen of the mystery,
> In which the heavy and the weary weight
> Of all this unintelligible world,
> Is lightened:—that serene and blessed mood,
> In which the affections gently lead us on,—
> Until, the breath of this corporeal frame
> And even the motion of our human blood
> Almost suspended, we are laid asleep
> In body, and become a living soul:
> While with an eye made quiet by the power
> Of harmony, and the deep power of joy,
> We see into the life of things.[34]

The shape and form of the world around us really comes down to a matter of composition and balance. By the mid-fifteenth century, we had swung too far toward seeing the eye of our minds. By the mid-eighteenth century, we had swung too far in the other direction, believing only what the eyes in our head tell us. But nothing lasts. We arrive at moments in human history where the two ways of seeing converge. I think Albrecht Dürer gave us such a moment in Gutenberg's wake. (Perhaps Galileo gave us another such moment in the near wake of Vesalius and Paré.)

But invention begins long before we reach such moments. Dürer was a great creative genius, no doubt. At the same time, he was also one of the products of the invention of printing with alphabetical, movable metal type. Our exploration of invention—of its texture and meaning—thus brings us to ask what the inventor really invents. We cannot just ask what came before the canonical inventor. We also need to ask what the entire arc of invention looks like. For while it is certainly true that the invention does not really begin with "the inventor," neither does it end at that near-mythological figure.

The invention of the printing press was poised to wreak yet another great transformation at the turn of the nineteenth century—a transformation about which I feel much more should be said than yet has been. Call it the democratization of literacy.

12

Fast Presses, Cheap Books, and Ghosts of Old Readers

W e're in a British parlor in 1806, listening in on a conversation. A tutor, whom we know only as Mrs. B, talks with two teenage pupils, Caroline and Emily. Their subject today is heat radiation (which, in accordance with the times, Mrs. B calls "free caloric"). She says to the young ladies:

> I should not quit the subject of free caloric without observing to you that different bodies, (or rather surfaces) possess the power of radiating it in very different degrees.
>
> Some very curious experiments have been made by Mr. Leslie on this subject, and it was for this purpose that he invented the differential thermometer; with its assistance he ascertained that black surfaces radiate most, glass next, and polished surfaces the least of all.

Emily pipes up, "Supposing these surfaces, of course, to be all of the same temperature?" Mrs. B replies:

> Undoubtedly. I will show you the very ingenious apparatus, by means of which he made these experiments. This cubical tin vessel or canister, has each of its sides externally covered with different materials; the one is simply blackened; the next is covered with white paper; the third with a pane of glass, and in the fourth the polished tin surface remains uncovered.[1]

Mrs. B fills the tin with hot water. Then she uses a focusing mirror to reflect the heat radiating from each side onto a thermometer. As she moves from the blackened surface to the paper, the glass, and finally the polished surface, each temperature reading is, in turn, lower.

Caroline, more inclined to play the bored, you-can't-fool-me role, is right on her. "The water within the vessel gradually cools, and the thermometer in consequence gradually falls." True, Mrs. B allows, but then she turns back to the blackened surface. The thermometer again gives a higher reading. In the end, the girls are convinced that the shiny surface is the worst radiator and the blackened one is best.

Emily, the more incisive and willing of the two, jumps in again: "According to these experiments, light-coloured dresses, in cold weather, should keep us warmer than black clothes, since the latter radiate so much more than the former." That too might be true, Mrs. B allows, if one were standing still in the shade. However, another property of black surfaces is that they absorb more of the caloric in the sunshine that falls upon them. Therefore, it is best to wear black in cold weather.

Now the plot thickens: Caroline and Emily begin posing thought models and arguing with each other. Gradually, they expose questions about the role of thermal equilibrium and energy conservation. This conversation is not only very sophisticated for the turn of the nineteenth century but also prescient—it catches all the right themes that mark the evolution of modern thermodynamics, just as the subject was finding its footing.

This scene is to be found early in the 1813 edition of a book titled *Conversations on Chemistry,* a book first published in 1806. The author is Jane Haldimand Marcet. Although these early editions of the book list no author, she begins her preface thus:

> In venturing to offer to the public, and more particularly to the female sex, an Introduction to Chemistry, the author, herself a woman, conceives that some explanation may be required; and she feels it the more necessary to apologize for the present undertaking, as her knowledge of the subject is but recent, and as she can have no real claims to the title of chemist.[2]

In fact, she had little to apologize for. And this book was only the first in Marcet's long series of such texts. The characters, Emily, Caroline, and Mrs. B, are fictitious, but they have become very real to me. Like Harry Potter or Nancy Drew, they appear in book after book. Marcet brings them back to life in *Conversations on Natural Philosophy, Conversations*

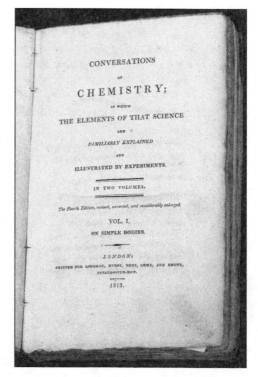

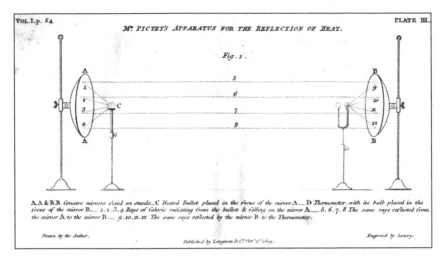

The mirror reflection system with which the radiation of caloric could be magnified and measured. In the experiment that Marcet describes, the heated bullet, C, would be replaced with the side of the water-filled tin, and mirror A-A would be eliminated.

on Political Economics, Conversations on Vegetable Physiology, and a host of *Conversations* on other less academic/scientific topics.

Marcet's work reflects two related changes that were taking place as the world emerged from eighteenth-century revolution. For one thing, attitudes toward invention were shifting in such a way as to bring about a dramatic change in our expectation of invention—the change described in Chapter 8. And learning was being democratized. (Indeed, Marcet was at the center of a significant underground of technical and scientific women.)

These changes were interwoven, for they had a common root. Both rode upon an area of invention that was as dramatic in its effects as the printing press itself had been 350 years earlier. Radical improvements in book production were afoot, and they would once again bring radical change with them.

The new output of printed matter, and new means for putting it to use began in Europe—especially England—but they found particularly fertile soil in America. Think, for example, about famous American inventors, later in the nineteenth century. Orville and Wilbur Wright, for example, were not university-educated, nor did they have access to an industrial organization. Yet their success in building a flying machine rested upon a deep understanding of the work in aerial locomotion that had gone before them. How did they know so much? The answer lies in the profoundly direct way that our dispersed population in America dealt with the problem of educating itself: they read.

Being self-taught was nothing new of itself. It had once been the norm more than the exception. But now literature penetrated society as never before. A great part of America's unschooled nineteenth-century population was remarkably well read. One need only listen to readings of Civil War soldiers' letters to know how well America had tuned its ear to the rhythm of language—to the style and content of books. If this print revolution was born in Europe, America would immediately join it and become its greatest beneficiary. What such English visionaries as Marcet were doing with those books was like water in the desert in our spread-out nation, raw and diffuse as it was.

To better see how that had taken place, I find it useful to close my eyes and remember those times I've watched large newspaper presses at work. Huge rolls of paper, weighing over a ton, feed a machine that stretches far across the print room floor. Paper moves far faster than the eye can follow as the press prints, cuts, and folds it. Those immense machines touch something beyond mere sight and sound.

Compare that with a Gutenberg press with two printers finishing a few hundred sheets a day, and the magnitude of the early-nineteenth-century printing transformation is clearer. Printers still used wooden hand presses during the eighteenth-century Industrial Revolution. Only after that did bookmaking begin to change. Inventors began building heavier, more stable frames. They began adding clever mechanisms to speed the printers' throughput.[3]

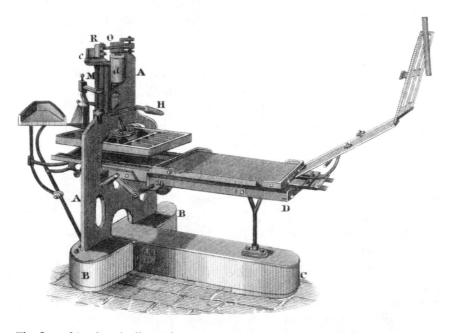

The first of Stanhope's all-metal presses.

Then, around 1800, the Earl of Stanhope changed the game. He built an all-metal press in which the operator used compound levers to drive a conventional screw mechanism. It imposed an increasing pressure that was highest right at the end of the printing stroke. Printers who once had to impress each half of one side of the paper separately could now, with far less effort, print a whole sheet in one pull of their lever.

The Stanhope press went through a variety of improvements and soon spawned more advanced all-metal designs. One of the notable ones was the Columbian press, built by George Clymer during the War of 1812, in Philadelphia. It was lighter than Stanhope's, and it included a major conceptual leap.

Instead of improving upon the old screw press, Clymer eliminated the screw drive entirely. He magnified the force of the printer's arms by means of a very sophisticated three-dimensional system of levers. From then on, it would be he, not Stanhope, whom others would build upon when they tried to improve the hand press.

Clymer cast the counterweight, used to lift the platen after each impression, in the form of a dramatic iron American eagle. Print historian James Moran tells how, as Columbian presses went international, other nations replaced the eagle with a globe, a lamp, or a lion. The Columbian press shown to the right is from the 1832 *Edinburgh Encyclopaedia*. Here, the American eagle is shown replaced with a simple round object, neutrally designated only with the letter *W*, for weight.

Clymer's version of the counterweight, **W**

These new iron presses provided a far more stable platform on which to add other features— means for loading the next sheet of paper, and re-inking the type, more quickly. While they made it much easier to produce high-quality printed matter, they retained the most serious limitation of the old wooden presses: They still could print

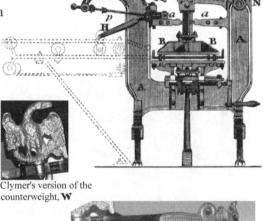

Clymer's Columbian press. Photos show details of Clymer's eagle counterweight and the complex 3-dimensional lever system.

only one page at a time. When we visualize that huge newspaper press, the old hand presses seem intolerably slow. They would clearly have to give way to some means for printing a steady flow of paper as it moved through some form of rollers. Moran tells of an idea offered by Italian inventor Faustus Verantius in 1616.[4] Verantius proposed rolling a large stone wheel over paper lying on inked type.

It might seem odd that we had to wait until 1790 before British inventor William Nicholson patented such a rotary printing system. Nicholson proposed using rollers both to ink the type, which itself lay on another roller, and then use the type roller to do the printing. Six years later an American, Apollos Kinsley, proposed his own rotary printing arrangement.

Nicholson never built a prototype, and Kinsley had too little mechanical infrastructure in our new country to succeed in making anything so complex. No doubt complexity is precisely what delayed such a machine until the nineteenth century. After more attempts by American and German inventors, German inventor Friedrich Koenig began developing such a machine in 1810. When the inking mechanism on his first machine failed to work as it should, he redesigned it and had a marketable press in 1811.

His steam-powered roller press could now print sheets of paper as they flowed by, on a flat platen. By 1814, he had figured out how to use a mating roller to press the type roller against

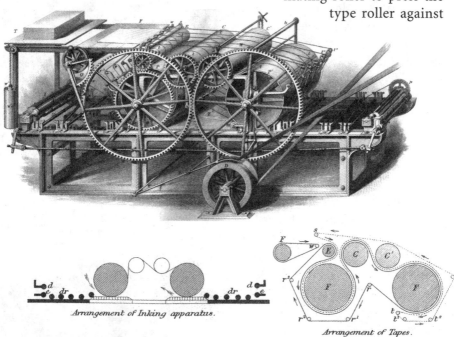

Above: A Koenig double roller press, manufactured by Applegath and Cowper in London before 1827. *Below*: The inking scheme, and the movement of paper, in this press.

the paper. Koenig also invented means for printing on both sides of paper. Think about that for a moment; just imagine the difficulty of doing two separate printings, front and back, on a moving sheet of paper, and placing each accurately on the sheet.

By the late 1820s those complex cylinder machines were major providers of printed matter. The printed word would now begin reaching the general public in ways that would transform the world and transform knowledge. By the way, Koenig roller presses still printed single sheets of paper. A variation on the roller press was developed later in the nineteenth century. It was the so-called web-fed press, which was fed by huge rolls of paper. A web-fed machine was used to print the newspaper we paused to watch a moment ago.

The Columbian press, along with its kin, would survive for a long time. It took a steam engine to run the new roller presses, and that high technology could be hard to come by. Iron hand presses remained in wide use until we had electric motors, also late in the nineteenth century.

These high technologies had yet to touch our 1813 edition of Jane Marcet's chemistry text. That particular book was still largely a product of the eighteenth-century bookmaking technology. Its format is *duodecimo.* That means twelve pages were printed on either side of a sheet of paper roughly twice as large (four times the area) of one of our letter-size sheets. Each of these large sheets was printed on both sides, then folded in half twice, then in thirds, to yield a twelve-leaf, twenty-four-page signature in which each page measures 7½ by 4½ inches.

These signatures were sold in uncut stacks to a customer who would normally take them to a bookbinder to be sewn together and bound with a proper cover. In this case, the folding was still being done by hand, and the pages are rough and uneven. They were printed on one of the new iron hand presses, but book demand was now speeding production, even if the presses themselves could not operate much faster than before. Haste

would have to accomplish what machinery could not yet accomplish. So prices fell as quantity won out over quality. People who previously could not afford books now could.

What happened next is best told to us by ghosts of former owners of these new books. And ghosts do whisper to us from their pages. Let us begin by meeting two owners of Marcet's chemistry book—one well-to-do, the other not. First, the well-to-do owner of this 1813 two-volume set.

A woman named Mary Anne Howley has signed both volumes. She was the daughter of William Howley, who later became archbishop of Canterbury. Her birth date is uncertain; it was probably around 1806. In 1825 she married the eighth baronet of Beaumont, and she died nine years later, not yet thirty years old.[5] We find very little written about Mary Anne Howley. A letter from her husband (a noted patron of the arts) to the famous painter John Constable says that they won't be able to meet since Lady Beaumont has suffered "a violent relapse of the prevailing Epidemic." That was a year before her death, and five years after she gave birth to the ninth baronet.

Marcet's chemistry book came out when Howley was around seven, but her signatures have the speed and fluency of someone older. Most likely she read the book as one of the teenagers who were Marcet's in-

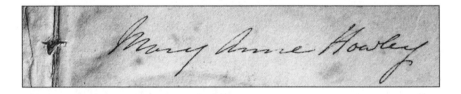

tended audience. Only the first two hundred pages of the first volume have been cut; someone has yet to take a letter opener to the remaining folded signatures. One has the feeling that Mary Anne Howley became distracted about the time Marcet was explaining hydrogen.

But she appears to have finished reading the part on heat. And here is another teaser: the noted artist who painted Archbishop Howley's portrait was Charles Leslie, son of an American instrument maker and friend of Ben Franklin.[6] Marcet, who had studied art and who did her own illustrations, talks about an ingenious experiment on free caloric (radiant heat) done by a Mr. Leslie. The whisper of centuries-old ghosts is indistinct, of course. So I may only hope that, with Marcet's connection to the art world, that might have been Charles Leslie's father, whose friend Ben Franklin also had experimented with radiant heat transfer.

Science and art were undergoing huge changes at the time, and determined women (such as Marcet) were asserting a new role for their gender. If Mary Anne Howley failed to emerge from the shadows of history just then, others would. For others, these books were agents of transformation, while for Howley, we have the feeling that this was only a talisman of hope for a better future.

Our second reader of Marcet's chemistry book was a boy whose identity I shall withhold for a moment. His prospects were far less promising than Howley's, but his story has a much happier outcome. He was born poor in 1791, the son of an out-of-work blacksmith. Worse yet, he spoke and wrote with great difficulty, his memory played tricks on him, and he did poorly in the symbolic language of mathematics. These are all symptoms of what today we call dyslexia.[7] He was, however, gifted with an uncanny ability to visualize—to see things whole.

The boy's break came at the age of thirteen. He had been apprenticed to a London bookbinder, and what happened next is best told in a letter that he wrote as an old man, fifty-four years later. It is addressed to his friend Auguste-Arthur de la Rive, the French scientist:

> My dear Friend,—Your subject interests me deeply every way; for Mrs. Marcet was a good friend to me, as she must have been to many of the human race. I entered the shop of a bookseller and bookbinder at the age of 13, in the year 1804, remained there eight years, and during the chief part of the time bound books. Now it was in those books, in the hours after work, that I found the beginning of my philosophy. There were two that especially helped, the "Encyclopaedia Britannica," from which I gained my first notions of electricity, and Mrs. Marcet's "Conversations on Chemistry," which gave me my foundation in that science.
>
> Do not suppose that I was a very deep thinker, or was marked as a precocious person. I was a very lively imaginative person, and could believe in the "Arabian Nights" as easily as in the "Encyclopaedia." But facts were important to me, and saved me. I could trust a fact, and always cross-examined an assertion. So when I questioned Mrs. Marcet's book by such little experiments as I could find means to perform, and found it true to the facts as I could understand them, I felt that I had got hold of an anchor in chemical Knowledge, and clung fast to it. Thence my deep veneration for Mrs. Marcet—first as one who had conferred great personal good and pleasure on me; and then as one able to convey the truth and principle of those boundless fields of knowledge which concern natural things, to the young untaught, and inquiring mind.[8]

This remarkable letter is signed by none other than Michael Faraday, one of the greatest scientists of the nineteenth century. Not only does Faraday pay grateful homage to our little-known textbook writer; he also lays out the dynamic of learning as it would mark his new print-driven century.

During the last year of Faraday's apprenticeship, Sir Humphry Davy was giving a series of dazzling public lectures in chemistry. Faraday attended them and took careful notes. He bound the notes in a book and sent it to Davy. In 1813 Davy hired him as an assistant, and only eleven years later, the Royal Society made Faraday a fellow for work on electromagnetism.

In one early experiment, Faraday created the predecessor of the electric motor when he used an electric field to spin a magnet. He went on to explain electrolysis, dielectric constants, and induction. He set the stage for Maxwell's field theory a little later in the century. The quiet Michael Faraday had found his voice, literally as well as figuratively, for his lecture demonstrations were spellbinding.

H. Adlard. sc.

Michael Faraday.

He was part of a gentle, offbeat, fundamentalist sect called the Sandemanians. They believed, among other things, in forming loving communities. So while other scientists waged scientific priority wars, Faraday created science lectures for young people. He used science to express his belief in the unity of nature. The agnostic physicist John Tyndall once remarked that Faraday drank from a fount on Sunday that refreshed his soul for a week. He must have, for this genius educated in a bookbindery—this lover of children and nature, this reader of books, this electric inventor—had near mystic means for seeing through to the very core of things.

In any case, the crowning irony appears in an 1833 edition of Marcet's chemistry.[9] The editor has added a version of the experiment in which Faraday anticipated the electric motor. In that surpassing bit of irony, the book now carried in it the fruit of her first edition.

The book had also changed in other ways by then. It was no longer handmade in Mr. Reibau's bindery on Blandford Street in London by a teenage Michael Faraday, for young ladies who were barred from attending universities. Now it was being churned off the new rotary presses, to be read by boys and girls alike, in schools and outside them.

A particular copy of another Marcet book provides a virtual laboratory in which to trace the complex symbiosis among the new fast presses and the creators of everyman's literature that she pioneered. It is a much later edition of her *Conversations on Natural Philosophy*. This book was first published in London in 1819.[10] The umbrella of natural philosophy then included almost any science. The term swept in astronomy, geography, biology, zoology—whatever an author fancied. In this book, Mrs. B, Emily, and Caroline talk about materials and motion, Newton's laws, hydraulics, and heat, light, and electricity.

Fifteen years after Marcet first published her natural philosophy, we find a copy of her 1834 American edition. As we look at it closely, we can

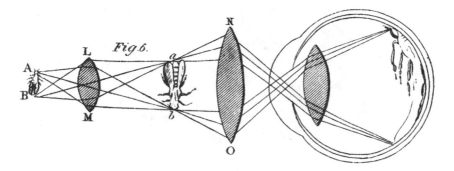

Marcet shows how one's eye sees a bug through a compound microscope.

trace the entire story of the early-nineteenth-century print revolution in microcosm. We shall look closely at six aspects of this particular book: its condition, its ownership, its authorship, the purpose of this edition, the quality of its paper, and how it was produced.

Condition. The book's condition can be described only as wretched. It has been used, stained, and worn down to the nub.

Ownership. In the front of the book we find Joseph L. Whittenburg's ink-blotty signature, dated May 13, 1838. He used a poorly cut quill pen that splayed erratically as he tried to write with it. Seventeen-year-old Whittenburg represents himself as a farmer whenever he shows up in subsequent census records.

He was born in Missouri and lived for some time in Alabama. He died in Texas in 1903 or 1905. Wherever he was when he got this book, he was still living in a remote and harsh land.[11] He signed the book several times more—in 1846, 1848, and 1850. Was he making note of the times he went back to refresh his understanding, or just reasserting his ownership? Alas, we cannot ask him, for he left even fewer tracks in the record of history than Mary Anne Howley did.

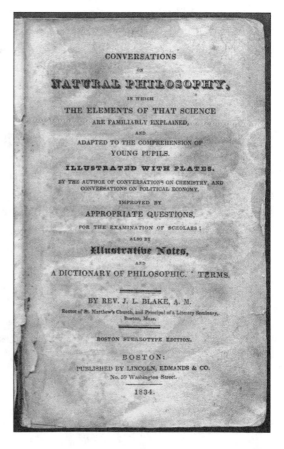

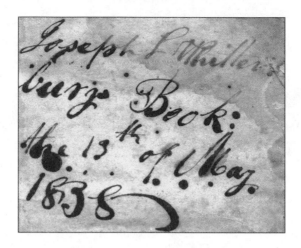

Authorship. Like the reader of this book, the name that appears on the title page is also male. It is that of the Rev. J. L. Blake. Marcet's pedagogy had matured, and *Conversations on Natural Philosophy* had a solidity that would keep it before a much larger public than she had originally envisioned. From the beginning, the book had served all walks of life, not just middle-class young ladies. Early in the book's history, edited editions appeared, and they continued to do so over a forty-year period. As with her chemistry text, the names of the many male editors would often cloak the fact that a woman was the author. J. L. Blake was only one of many such editors.

The book's purpose. At first glance, Blake's presence on the title page might seem like blatant plagiarism. Yet it should be seen in the framework of the times. Blake identifies the writer not as Marcet but as "The author of Conversations on Chemistry and Conversations on Political Economy." In his preface (written in 1824 for his first edited edition of the book) he acknowledges her by name and speaks respectfully of her.

He says that he had used her original text for three years in the seminary where he teaches, but had finally decided that his classroom students needed more structure than Mrs. Marcet provided. Her book was, after all, meant for self-study. Blake has added study questions and a "dictionary of philosophical terms" (what we would call a nomenclature section). In that we see a diminishment of the fierce independence that was driving the individual readers of such literature.

But was Blake's action sexist? Well, of course it was by our standards. But compared with the way women were generally treated, he is being downright open-handed. I am more bothered by J. L. Comstock, who was first content to add homework questions to Marcet's book for use in American schools, but who then wrote his own text.[12]

Comstock changed the tone of Marcet's book far more than he changed the content. The easy, conversational style became authoritative and didactic. The 1848 edition of Comstock that I had access to had belonged to a boy at the exclusive Phillips Andover Academy. Marcet was still shaping America, but now her illustrations have been made stiffer and less elegant. The graceful style of the conversation among Mrs. B, Emily, and Caroline has gone dry.

The paper it was printed on. The paper is fragile and badly foxed, with variable brown stains throughout it. Foxing does not occur in wood pulp paper, but does in the rag papers made from the late eighteenth century onward. It was at its worst in the years during which these later editions of Marcet's books were being made. Though foxing is poorly understood, it is probably the result of oxides of the iron left in rag paper by certain kinds of processing.

High-quality paper is made from the same vegetable fibers that are used in textiles. In Europe, cotton and linen rags were favorite sources. That required a peculiar balance: once the rate at which a population cast off its worn cotton garments was exceeded by the rate at which it demanded paper, it was in trouble.[13]

That is exactly what happened. Even without fast presses, increasing quantities of paper had been consumed during the eighteenth century. It was used in a huge upwelling of newspapers, broadsides, the first magazines, and gift wrapping, as well as for books. Soldiers used paper to wrap black powder and lead musket balls. When the Continental Army drove the English out of Philadelphia in 1777, they reentered the town badly in need of paper for their ammunition. They finally found 2,500 copies of a sermon on "defensive war" in Benjamin Franklin's print shop and put them to use in their own defensive war.

Housewives all over Europe and America came under intense pressure to save and recycle every scrap of old clothing for use in making paper. Since paper cannot be made from animal fibers, the British went so far as

to pass a law requiring that only wool be used for burial garments. That freed 125 tons of cotton rags per year for English papermaking—a pitiful drop in the bucket. A replacement paper fiber had to be found.

The French scientist, René Réaumur, came up with what would prove to be the eventual solution in 1719. Look at the North American wasp, the so-called paper wasp, he said. They make fine paper for their nests by chewing up wood and exuding it. If they can do that, why can't we?

Wasp paper.

Réaumur did not actually make paper, but the idea stayed alive until a German clergyman, Jacob Schaffer, picked up on it. Between 1765 and 1771, he wrote a huge treatise on making paper from alternative fibers. He included actual paper samples that he had taken from wasp nests and ones that he had made directly from various woods.

But rag-based paper was too good to give up without resistance. Paper-makers limped along, stretching their rag slurries by adding straw, then grass. They tried to repair the color and quality and to chemically enhance the process. That is when paper quality began a long, slow decline.

Not until the 1850s did papermakers break down and begin using wood pulp. The first paper they made from wood was simply awful. The chemical processes of wood pulp papermaking remained crude until the late 1800s. Some books from the 1870s are so acid-eaten and brittle as to crumble under the slightest touch. Wood pulp paper subsequently began improving. Today, it varies from shabby to excellent.

This 1834 version of Marcet's *Natural Philosophy* predated any use of wood. The paper in it is variable, as was typical in such books. Different batches of paper were used for the text and for the illustrations, which are bound in the back of the book. The pages for the illustrations show much more serious foxing than the text; but, despite the variable discoloration, the cotton rag paper remains generally tough and flexible throughout.

The book's production. We read the following words on the title page: "Boston stereotype edition." Our word *stereotype,* to cast a person in a preset mold, comes from a copying process invented in 1725 by Scottish goldsmith William Ged.[14] Ged found that he could greatly speed the printing of a long run of pages by setting the page of type, then making a papier-mâché mold of that entire page. He used that mold to cast multiple copies of the page in lead. That eliminated the tedious task of setting multiple plates of type for use in multiple presses. He thus liter-ally invented means for "stereotyping" the plate of type.

Here we find ourselves in one of those nasty situations that often alters the course of invention. English printing was still held in lower esteem than that on the Continent. Paper in England had generally been of poorer quality and the printing less elegant than that in Europe. Five years before Ged's invention, a printer named Bowyer spotted some fine lettering done by a young brasscutter who engraved locks and barrels on fancy guns. His name was William Caslon—not to be confused with William Caxton (Chapter 10). Bowyer asked Caslon if he could cut type.

Caslon had never even seen type being cut, but he had artistic skill, so Bowyer bet on him. He handed Caslon £500 and told him to set up a type foundry. That was a good move; Caslon soon set a new standard for English type. He modeled most of his fonts on Dutch type, but his refinements created a clean, uniquely English typography.

Caslon ABCDEF

Caslon was the archetype of the superb professional—modest, able, reliable. But Caslon unfurled a different set of colors when Ged invented stereotyping and the old met the new. He sneered at Ged's process, boasting that he could duplicate plates just as fast. "I'll bet you £50 you can't!" Ged shot back. So each went off with a page of type. Ged made three copies before Caslon even got started.

Ged won the bet, and he won a contract to print books for Cambridge University, but now Caslon was angry. He planted saboteurs in Ged's shop. They ruined the work, ruined Ged's business, and ruined Ged. Stereotyping vanished for eighty years. And after Caslon died, his lovely typography fell out of use until it was revived in the twentieth century.

Ged was vindicated much sooner. His stereotyping returned in the early nineteenth century as a perfect part of the movement toward faster production of printed material. I still carry the strong memory of my father taking me down to the newspaper where he worked during the 1930s. I would scavenge lead scraps from their stereotyping process and melt them down to cast into my own stereotypical lead soldiers. For the late nineteenth century had produced the Linotype machine, used to produce stereotype plates directly from a typist's keyboard. Like the web-fed roller, Linotype had been added to those huge newspaper presses we talked about earlier.

How differently printing might have developed had that senseless competition between Caslon and Ged not taken place. But stereotyping was commonplace by 1834, and it had been used to make Joseph Whittenburg's inexpensive textbook. This book now has a cheap factory binding and its edges have been machine-cut. They are smooth and uniform. This is a book in the mode of our times, no longer in that of Jane Marcet's.

As the new production books came off the presses, perhaps the most prolific creator of cheap educational literature was a Cambridge-educated mathematician, the Rev. Dionysius Lardner. He generated all kinds of handbooks and textbooks during the early to mid-nineteenth century. (Several of his illustrations are reproduced in Chapter 5.) Lardner's stable of ghostwriters and co-authors included some very distinguished writers. Mary Shelley, author of *Frankenstein,* was among them.

Lardner provided a whole array of books that included the term *natural philosophy* in their titles. His treatments of machinery were more extensive than Marcet's, but his organization of the subject again echoes hers. Lardner promises to provide the background in natural philosophy that "is expected in all well educated persons."

One of his readers was on his way to being among the most educated people on this planet. He was fifteen-year-old William James, who, in 1858, was studying an 1856 edition of Lardner's *Hand-Book of Natural Philosophy*.[15] James was no passive reader. In the back of his copy, he has penciled in his own graphical construction of the catenary curve—the pendent shape of a suspension bridge chain—done in accordance with Lardner's mathematics.

After the Civil War, Marcet's natural philosophy remained in print, but the many copycat textbook writers no longer included descriptions of machinery in the subject. Machines dropped out of liberal education, and the focus shifted to principles. Those principles became a specialty with a new name: *natural philosophy* was now replaced with the word *physics* (from the French *physique*).

Even then, however, Marcet's subject layout remained. It is apparent in physics texts today. But when the name of natural philosophy became physics, it ceased to lie at the center of liberal education, where Marcet knew it belonged.

Since women were not admitted to college, she had set out to create a home liberal education for them. Before she was done, she'd provided courses in chemistry, natural philosophy, economics, botany, and geography. Today, however, we barely remember that natural philosophy belongs in the core of a liberal education, and we completely forget that such an education must include the machines we live with every day. Today we might well find Marcet's ideas to be very provocative as America falls behind in technical and scientific education.

So a bright moment lost a part of its luster. Cheap text material shifted in form as it proliferated. Others picked up on Marcet's pioneering work as she lingered, ghostlike, in the background. Let us look for a moment at just one of those lingering afterechoes of Jane Marcet.

An unexpected article by Theodore Sterling is titled "Marcet's Apparatus." The article turns up in a modern journal on collecting old instruments.[16] The Marcet who developed this apparatus is not Jane but her son, François. The device not only has his mother's fingerprints all over it but also carries some very interesting hints as to Jane Marcet's larger influence upon the evolving social fabric of the later nineteenth century.

Marcet's Apparatus.

From an 1856 chemistry text.

The purpose of the instrument is to measure the vapor pressure of water. A small spherical boiler is fitted with a thermometer and a long mercury manometer. The boiler is half filled with water, then heated. The manometer and thermometer give a paired sequence of pressures and matching boiling-point temperatures as the sphere is heated. The instrument is limited by the height of the manometer. If it is 2½ feet high, it will provide a sequence of boiling-point measurements up to twice atmospheric pressure. If it is 5 feet high, it will give results up to three atmospheres, and so on.

Sterling estimates that François Marcet built this device around 1826, when he was only twenty-three. If that date is correct, François Arago of the French Academy of Sciences had done the same experiment three years earlier. Arago mounted sections of pipe up the side of a church tower and managed to measure water vapor pressures to twenty-four atmospheres. Sterling tries to determine whether Arago or Marcet originated the idea, and decides Arago was probably first.

Born in England, where he went by the name of Francis, François Marcet was a toddler when his mother wrote *Conversations on Chemistry* there. He was living in Switzerland when he built the device. Since Jane Marcet's husband, Alexander, was a noted Swiss physician, the family spent time in both countries.

Jane Marcet had served as the very effective focal point for other women scientists, since she was remarkably well connected both in England and on the Continent. Arago was one of her European scientific friends. That apparatus was almost certainly the fruit of shared ideas, and its significance reached beyond measuring water's boiling points. For both Jane Marcet and François Arago used technology as a vehicle for social reform. Both had highly honed senses of social responsibility. Marcet did much to take women out from behind their "crinoline curtain," while Arago strove to bring the Industrial Revolution to France.

An important part of Arago's mission was getting the French to adopt the new English steam engines. James Watt had been among the first to measure vapor pressures. Now both Arago and Jane Marcet's young son worked on improving that basic and badly needed engineering information.

In 1872, Francis Marcet (now sixty-nine years old) edited a fourteenth edition of his mother's natural philosophy text. By then steam engines were running at far higher pressures than those for which Marcet's, or even Arago's, apparatus could provide data, and those engines had transformed us. But, like his mother's book, Francis Marcet's apparatus was still serving as a part of college instruction.

In any case, we read the impact of the fast presses in specific copies of these books. Ghosts speak to us from their pages because ghosts are more plenteous in them than they are in their more respectable fine-press kin. One did not carry a book worth a month's pay on a Sunday picnic. And yet, as books became more rough-and-ready, readers became more deeply invested in them, not less. The amount of writing in margins increases. More often than not, bindings have cracked and pages are well thumbed. There is room for ghosts in these old mansions.

Marginalia is nothing new. However, readers from every walk of life reveal themselves to us with immediacy and intimacy in almost any copy of these books that we pick up. Let us therefore meet one last ghost of another time before we move on to other things.

His name was Joseph Foster, and he turns up in an old mathematics text, *Elements of Geometry and Trigonometry,* published in 1855.[17] Although

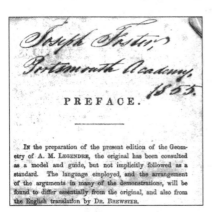

the author's name was Charles Davies, the name Legendre is stamped on its calf cover. The great French mathematician Legendre had long since died. However, by saying that his work is based upon Legendre's, Davies attempts to add luster to the work. Davies wrote other books and invoked other famous French mathematicians. This is, in fact, an advanced high school trig text. Foster was then a fourteen-year-old schoolboy in New Hampshire's Portsmouth Academy who had received the book when it was new. He inscribed it in a flowing young hand.

Like Joseph Whittenburg, Joseph Foster has signed his book again, in another place. But now the schoolboy flourish is gone, and what it says this time catches us off guard: "Joseph Foster, U.S. Steamer Augusta, off Charleston, S.C., 17 April, 1863." Foster had joined the Union Navy and was now part of the Charleston blockade in the Civil War.

He shipped out on the *Augusta,* a steam-powered, paddle-driven Union gunboat. America had, for the first time, sent steam off to war, and had sent Foster off with it. Ten days before Foster inscribed his book the second time, the *Augusta* joined Admiral duPont's attack on Charleston. That attack had failed miserably. The Union Navy retreated into the bay, and duPont was relieved of his command.

Now the ghost of Foster reaches across the years to tell us he was there—that he saw men die. My old text on geometry and trigonometry was there with him. So I reached back to look for Foster and caught up with him in a library in Portsmouth, New Hampshire.[18] He had been twenty-two years old during the defeat at Charleston—eight years from the schoolboy days when he first got this book, and he had carried it into war with him.

Foster served on three of the new Union gunboats during the war: the *Augusta,* a 1,310-ton sidewheel steamer; the *Acacia,* a 300-ton screw steamer; and the *Commodore MacDonough,* a 532-ton sidewheel ferry.[19] The *Commodore MacDonough* sank on August 23, 1865, as it was being towed from Port Royal back to New York after the war. Only Foster managed to save any personal effects, and all he saved were some books and papers, including this old geometry book.

Foster continued to serve in the Navy in a variety of storekeeper and paymaster roles until he retired in 1902 at the rank of rear admiral. By then, he had lived all the way through the Spanish-American War, but he had seen no more combat. Back in Portsmouth, he ended his life writing histories and biographies, which may now be found in the Portsmouth Athenaeum. He died on May 19, 1930, three months before I was born. His old school became the city library in Portsmouth, while his old book found its way to my shelf.

It is clear from Foster's marginal notes that he bogged down after studying plane geometry. Yet he carried this book through the war, and I think I know why. It was a young man's time out of war—a place of retreat into a clean and beautiful place. My old book had been Foster's handhold on sanity in a world that had, for a season, lost its wits.

The sudden ubiquity of books had cast reading into a new relationship with learning. The organization and character of education would have to change as a result. Today, we look at all those computers upon our desks and ask one another how we will go about reinventing education. Perhaps it will help to look back at the way education was reinvented in the wake of the fast presses.

Part **IV**
Views Through a Wider Lens

13

Inventing Education
The Great Equalizer

T o learn is to incur surprise—I mean really learning, not just re-freshing our memory or adding a new fact. And to invent is to bestow surprise—I mean really inventing, not just innovating upon what others have done. Learning and inventing can feel a great deal alike for just that reason. Both surprise us when we seriously undertake them.

It should be no surprise that those twin themes of learning and inventing are woven through the life of one of the most surprising people to emerge from the nineteenth century. He was Peter Cooper, born in New York City on February 12, 1791, a mere six months after the newly formed United States Patent Board had granted the first U.S. patent. He died ninety-two years later, on April 4, 1883, a scant fifty-seven days before the opening of the Brooklyn Bridge.[1] He ended his life in an America that had been transformed in many dimensions and beyond recognition.

Cooper had been a formidable agent in bringing those transformations about. They were woven through, and had defined, his life—a life that touched so many themes we have been considering: the trajectory of invention, railways, egalitarianism, and even flight. But the red thread through it all was his fascination with a way of learning, and his sense that he was obliged to capture that way by re-creating education.

Peter Cooper, from the 1897 *Encyclopaedia Britannica* supplement.

The processes of education have had to be reinvented as surely as industrial processes have. Just as the design of automobiles changed after the invention of the assembly line, methods of education had to be reinvented after each major change in the availability of information—the

— 217 —

ninth-century increase in literacy, the thirteenth-century spread of manuscript copying scriptoria, and the fifteenth-century development of the printing press. In Cooper's time, the fast presses of the early nineteenth century were demanding yet another reinvention of education, and today we ask how we should be retooling education in the midst of the Internet revolution.

Peter Cooper's story is particularly poignant since while he himself was uneducated in any formal sense, he very clearly understood the changing fabric of the world around him and the role being played by reading and learning. His is a story laced with contradiction: Cooper was a great inventor, yet gifted with a talent for putting each successive invention behind him and moving on to something new. He was a wealthy industrialist, but he was also a creative social reformer and an influential egalitarian.

Cooper was the son of an idealistic Methodist minister who had kept his peripatetic family in constant financial trouble. As a young boy, Peter demonstrated an astonishing creative fearlessness and a complete lack of concern for life and limb. He was lucky to emerge only with many scars, and not seriously maimed or dead.

He was apprenticed to a coach maker as a teenager, and his genius for invention showed up early. During that period he invented a machine for shaping wheel hubs that was still in use when he died. As for education, he had only about a year of schooling, and he read what few books he had access to—the Bible and Pope's *Essay on Man*. Throughout his life he, like Michael Faraday, was propelled by a powerful yet liberal religious conviction. During his apprenticeships he lamented the fact that education was not available at night for those who had to work by day. Two talents, however, saved him. One was an uncanny ability to draw content from conversation. The other was an equally astonishing ability to visualize mechanical processes.

While Cooper was still a teen, he devised a complex scheme for getting power out of ocean tides, built a model, and arranged to demonstrate it to Robert Fulton. Fulton gave him no response at all, only stony silence. That event appears to have carved yet another scar into Cooper. He kept the model all his life—a dangling hope; an unhealed wound? We do not know. We do know that he came away more determined than ever to invent. He patented a musical cradle, a process for making salt, a rotary steam engine.

Then, in 1821, Cooper had an opportunity to buy a failing glue factory for a low price. He knew nothing about making glue, but he dove in to both redesign the factory and reinvent the processes by which glue was made. He was soon making improved calf's-foot gelatin and isinglass, as well as improved glue. He became wealthy in his thirties.

That was a pattern he would repeat. In 1828, with his glue factory now well established, Cooper invested in an iron-smelting operation in Baltimore. That proved to be a large, complex, and star-crossed venture, one piece of which was moving goods between Baltimore and New York.

The embryonic horse-drawn Baltimore and Ohio Railway would have to be the means, but it was clear to Cooper that it was a clumsy one. So he invented and built a steam locomotive—America's first steam locomotive, as it turned out. He named his engine the *Tom Thumb*. At least that's what he called it when he was not calling it *Teapot*. He intended the *Tom Thumb* to be only a demonstration, not a prototype. Still, in 1830 it managed to average 14 mph on a thirteen-mile run. His point was well made, and from this beginning America went on to overtake Great Britain's strong lead in railway technology.

Model of the *Tom Thumb* in the Museum of Science, Boston.

For other inventors, the *Tom Thumb* might have been the accomplishment to be carved upon a tombstone. But Cooper had created America's first locomotive almost as a throwaway. Another throwaway was less successful. He became interested in flight and designed an airship. The gasbag exploded as he was filling it, and a shard of flying glass did permanent damage to his left eye—one more scar upon a man who truly believed in nothing ventured, nothing gained.

Business flourished under Cooper's flair for far-out, farseeing projects. Along the way, he went far beyond just becoming wealthy; he became fabulously rich. And yet the story of Peter Cooper cannot fully be told without involving a second person—a kind of ying-yang mirror opposite who completes the picture.

He was Abram S. Hewitt, born in New York in 1822, the fifth son of a cabinetmaker. Hewitt, also an inventor and craftsman, went on to become a steel magnate to rival Carnegie, a five-term U.S. congressman, and finally mayor of New York City.

Yet Hewitt's biographer Allan Nevins begins by saying, "This book is concerned less with a single individual than with a family [of two branches] . . . the Coopers and the Hewitts: a family which spans American life from 1791 when Peter Cooper was born, until 1903, when Abram S. Hewitt died."[2]

Hewitt's and Cooper's lives were the warp and woof of a single fabric— Cooper was gifted with daring and vision, and Hewitt managed to gain a secondary-school education, then go to college and obtain a law degree at Columbia. One of Hewitt's many jobs as he worked his way through college was tutoring Cooper's son Edward at Columbia. Peter Cooper took a huge liking to the young Hewitt. He practically made a second son of him, as well as making him a business partner. By the time Hewitt married Cooper's daughter Amelia, he was the thirty-three-year-old executive of Cooper's immense steel business.

Cooper had gone into the steel business on his instincts; now Hewitt focused on modern developments in metallurgy. He was the first American to try out Bessemer's new process for making steel. When it failed to be all that he had hoped, he then led America in adopting the Siemens-Martin open-hearth steel converter. When Cyrus Field began work on the Atlantic Cable, Cooper was his prime backer, but Hewitt provided lobbying support in Washington, as well as telegraph wire.

Looking at Cooper and Hewitt, one might be tempted to ask which of the two should get the prize for the life best lived. Of course, the question itself is wrong! Whether by intent or by subconscious reflection, Hewitt finally weighed in on the matter less than two months after Cooper died, when he was asked to give the dedicatory address for the opening of the Brooklyn Bridge on May 31, 1883. As Nevins describes the speech, Hewitt might well have made the bridge into the metaphor for what was really on his mind. According to Nevins,

> engineers, he [Hewitt] said, are of two kinds, the creative and the constructive, and the creative engineer, like the poet, is born, not made. "If to the power to conceive is added the ability to execute, then we have one of those rare geniuses who not only give a decided impulse to civilization, but add new glory to humanity."[3]

Hewitt was ostensibly referring to the bridge's designer, John Roebling. But he beautifully captured the texture of his symbiosis with Cooper. For they were two people who each knew, on a gut level, that the other was his perfect complement, two people who together had surpassed even Roebling in the reach of accomplishment.

This is certainly a tale of wealth and influence, yet those aspects should not be misunderstood. Neither of the two was the rapacious nineteenth-century industrialist that we so often read about. Cooper, in particular, was fueled by idealism. He had been a passionate abolitionist, but not a Lincoln supporter. Early in life, he had been a Jacksonian populist, and

Abram S. Hewitt, from the 1897 *Encyclopaedia Britannica* supplement.

later in life, a staunch Democrat. Yet the Republican Lincoln entered Cooper and Hewitt's horizon through the door to one of Cooper's most intensely hands-on inventions—one that particularly captures my fancy.

It was New York's Cooper Union. Why Union? Because the college was meant to unite science and art. Indeed, it would have been named only *Union* if Cooper had had his way. The state legislature intervened and added his name. Cooper Union was meant to carry out Cooper's belief that "education be as free as water and air." He wrote, for example, of the new school that it should be "open and free to all who can bring a sertificate of good morrel character from parent guardian or employer."[4] (With almost no formal schooling, Cooper's spelling was just as inventive as his machines.) Cooper had enormous feeling for those who would otherwise not have the opportunity of going to school. He also placed a great emphasis on educating women so that they would have options beyond being trapped into early marriage. Still small and tuition-free, Cooper Union continues his mission to this day; if one has the talent to attend, one does so on a scholarship.

The Union had been on Cooper's mind for a long time. As early as 1838 he had bought a plot of land for it near present-day New York University, and he began building it in 1853, using a radical wrought-iron rolled-beam construction of his own devising. Its five-story height made the upper floors something of climb in a pre-elevator era, but Cooper's vision was at work again: he left space for the installation of elevators, at the time a technology not yet developed to the point where it could be used. He insisted on that over his architect's protests, and elevators run in those spaces today. (Cooper made those elevator spaces round, not square, on the basis that a round elevator would hold more people for the amount of material that went into it.)

Nowhere was the complementarity of Cooper and Hewitt more pronounced than it was in Cooper Union. Cooper's mind ran free, as he set out to redefine education with the help of architecture. He would put a garden and a bandstand on the roof, a lecture hall in the basement where it would not be disturbed by street noise, a museum on the third floor. He put a huge emphasis on showing rather than just telling. Hewitt, in charge of formulating curriculum and fitting it to the building, was Cooper's counterbalance, modulating the flow of Cooper's ideas and making them workable or nixing them.

When the school opened in 1858, the name Union was taking on metaphorical meaning, for the Civil War was looming. On February 27, 1860, the Young Men's Central Republican Union rented Cooper Union's main lecture theater for Abraham Lincoln, then a presidential hopeful, not yet a candidate. Cooper had been working, and he arrived late, just as William Cullen Bryant introduced Lincoln.

Lincoln started slowly, even awkwardly. But once the lawyer from Illinois got rolling, Cooper was riveted by his logic, intelligence and mastery of facts. The Cooper Union address was probably the turning point in Lincoln's presidential campaign. Cooper was a pro-South Democrat, but two factors turned him toward the Republican Lincoln: his abolitionism made him vigorously opposed to allowing the expansion of slavery into the territories, and he regarded the breakup of the Union as intolerable.

After Lincoln was elected, Cooper peppered him with advice. At one point, he wrote to Lincoln with the earnest suggestion that he could avert war if the North would simply buy all the slaves and free them, after which they could work for salaries. If buying off 4 million slaves sounds crazy, compare it with an 1860 cost of $4 billion and half a million lives lost in the war. Then it gains a great deal in sanity.

Cooper was no more successful in that suggestion than he had been with his flying machine or than he was in 1876 when, at the age of eighty-five, he ran for president on the third-party Greenback ticket. Samuel Tilden won the popular vote but lost the election to Rutherford B. Hayes. Cooper picked up only 0.9 percent of the votes cast. (Ten years later, Congressman Hewitt, with far more political savvy, took on Theodore Roosevelt in the race for mayor of New York City and beat him handily.)

But failure was part and parcel of Cooper's immensely risky and creative life. In the end, another Cooper idea had to undergo modification by Hewitt and others. When it did, it ultimately became a profoundly important contribution. It was that museum he had envisioned on the third floor of Cooper Union.

Instead of a museum, it emerged as a public reading room. At this point the public library was just being invented in America, and the Cooper Union reading room was an important step in that process of invention. Just a few years earlier, New York's Astor Library had opened to the public, but its mood was very different. It served the leisure classes and closed at five o'clock. Cooper insisted on his library being open to the public between 8:00 a.m. and 10:00 p.m. seven days a week so that working people could use it.

And use it they did! The Cooper Union library received over four hundred people per day during its first year. By 1880, that number had risen to two thousand per day—young and old, rich and poor. This was education as Cooper understood it, serving the hunger for knowledge and the hunger to read simultaneously.

Cooper's reading room was only a small step away from the uniquely American institution of the public lending (as well as reading) library. By placing it where it received such huge traffic, he did much to show America what libraries could be.

Inventing Education

You and I tend to see our libraries (like our automobiles and telephones) as having been there always. When we celebrate the ancient libraries, our mental picture is apt to be formed by the building down the street. Just how much change has gone on becomes clearer when we look at the rich history of libraries provided by the 1911 and 1970 editions of the *Encyclopaedia Britannica*.

The libraries that we know—those public places where the books that we jointly own are catalogued, maintained, and lent out to us— are not much older than automobiles or telephones. The idea of sharing our goods in such a way is, of course, pure socialism (of a kind that the capitalist Cooper would have applauded), and the library came into being with other socialistic institutions that arose in the mid-nineteenth century.

Free public lending libraries had emerged from time to time in America since the seventeenth century. The Massachusetts Bay Company provided a public collection of books as early as 1629. Salisbury, Connecticut, created a public reading room for children in 1803. Perhaps the oldest public library (in the sense we know it) is the one established in Peterborough, New Hampshire, in 1833. New Hampshire was the first state to establish general funding for public libraries and the first to make such an institution official. After it passed the necessary legislation in 1849, public lending libraries were here to stay.

The British *Encyclopaedia Britannica* makes no bones about crediting America with this remarkable institution, which was clearly part and parcel of the new movement toward American social reform. It is significant that 1850 was the same year the first nationwide labor union was formed, and considering all we have been talking about, it might be no surprise that it was the printers' union.

As the impetus toward the creation of libraries hurtled ahead, Peter Cooper's idea of a museum on the third floor of Cooper Union resurfaced in conjunction with that impetus. In 1897, Hewitt's three daughters, Amy, Eleanor, and Sarah, finally created a museum

RICHMOND LIBRARY ASSOCIATION.

No. 1281

RULES AND REGULATIONS.

1st.—Any person shall have the privilege of taking books from the Library for one year, by paying one dollar in advance, for six months, by paying fifty cents in advance.

2d—No person shall be entitled, under Rule 1st, to have more than one volume at a time from the Library.

3d.—Any person keeping a book from the Library for a longer period than two weeks, shall pay five cents for the third week and ten cents for every additional week the book is kept from the Library.

4th.—Any person wishing to take more than one volume from the Library at a time, can do so by paying five cents per week, for each additional volume so taken out for the first three weeks, and ten cents per volume for each additional week the book is kept from the Library. But in no case shall any person have more than three volumes from the Library at a time.

5th.—The Library will be open for the delivery and reception of books, from two to four o'clock every Saturday afternoon, except the Saturday immediately preceding the first MONDAY in JANUARY, when it will only be open for the reception of books; AND ALL BOOKS MUST BE RETURNED TO THE LIBRARY ON THAT DAY.

6th—The annual meeting of the Association for the choice of officers and other business will be held on the first MONDAY in JANUARY.

This notice, pasted in an 1872 edition of a science text, is probably representative of a date shortly thereafter. This library, while public, requires an annual fee of fifty cents for library privileges.

that was faithful to their grandfather's principles. Their Cooper-Hewitt National Design Museum in New York, located in a former Andrew Carnegie mansion, is a branch of the Smithsonian Institution and is quite active today. The museum combines artifact displays with descriptive literature, providing a space in which we can read books and notebooks about the artifacts before us. In a booklet about the museum, Eleanor said of her grandfather:

> Mr. Cooper, while planning his great building for the "education and recreation," arranged for one whole floor to be used as a Museum for the exhibition of mechanical devices and a Cosmorama, for the pleasure, instruction and enjoyment of all who could not visit foreign lands.[5]

She goes on to tell how donors appeared to support Cooper's revitalized idea, and she calls the project "a curious weaving of the web of Fate, where artistic and mechanical tradition appear to have accomplished a foreordained result."

Cooper's vision was still working itself out a half century later. Americans, who had become remarkably literate by the mid-nineteenth century, would read as no other people ever had. More institutions had followed the creation of public lending libraries, and they laid the foundation for what Eleanor Hewitt called an "artistic and mechanical tradition."

In 1862, for example, Congress had passed the Morrill Act, which mandated a grant of 30,000 acres of federal land, per congressional representative, to each state to be sold to provide an endowment for

> at least one college where the leading object shall be, without excluding other scientific and classical studies and including military tactics, to teach such branches of learning as are related to agriculture and the mechanic arts.

Here we clearly read America's determination that we would be a free people, with a liberal education—a truly *liberal* education, that is—one focused upon the agricultural and mechanical arts but not excluding the rest of what we see today as necessary for a healthy general education.

Great Britain's 1911 *Britannica* conceded that America already had ten thousand libraries of a thousand volumes or more. Today, I could stand in my yard and throw stones that would hit several private homes with holdings that large, and so could you, for we have reaped the fruit of having socialized our reading.

When I did my undergraduate engineering at what is now called Oregon State University, I would pass a wrought-iron fence on the way to class. The letters *OAC* were woven into the ironwork. The old name, Oregon Agricultural College, had been changed, but the impetus was the same. I was the beneficiary of a mid-nineteenth-century decision that we rank and file Americans would read and be educated. That decision was another product of the remarkable idea that you and I would *jointly* own far more books than any of us could own alone.

The Morrill Act encapsulated another impetus of the times—an idea that we may have allowed to grow rusty, as we've let technology separate itself from other branches of learning. Technology had been the means by which Great Britain had carried out social revolution in the late eighteenth century, and technology was now playing that role in America. Technology still lay at the heart of our invention of general education.

But those cheap old books, spawn of the new fast presses, were the vehicles for those revolutions. When we open them and meet the ghosts who occupy them, we can often trace the course of a social change as it builds upon the synergy between a mentally independent reader and a lucid book.

Take, for example, a book that was produced in the same way as Mary Ann Howley's natural philosophy text (Chapter 12). But, unlike hers, its pages have all been cut, and throughout it, pencil marks make it clear that the owner had digested it carefully. The book is a part of a series that the Rev. Dionysius Lardner called *The Cabinet Cyclopedia, Natural Philosophy*. It was written by one Captain Henry Kater, and the title is *A Treatise on Mechanics*.[6]

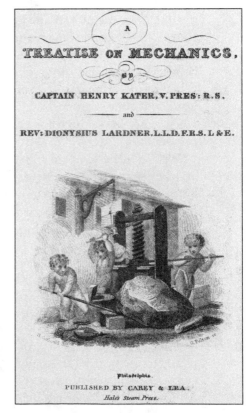

As we turn pages, we read about matter, and about forces that act upon matter. The book explains pulleys, gears, engines, and screws.

But more is going on here. Kater is also spinning a philosophy around the new atomic descriptions of matter, and his words obviously resonate with the book's young owner. He is a twenty-two-year-old student at the University of Pennsylvania named Richard Newton, and he has marked this startling sentence by Kater:[7]

Kater's title page includes an image of four forms of levers: cantilever beam, wedge, lag screw, and simple lever. Three cherub-like children dramatize the advantage they offer.

> Although we are unable by direct observation to prove the existence of constituent material atoms of determinate figure, yet there are many observable phenomena which render their existence in the highest degree probable, if not morally certain.[8]

Morally certain? Hardly the way anyone would talk about atoms today. But Kater was no rambling generalist.[9] He had started out as a creature of the old empire. He had studied some law as well as mathematics. Then he purchased a commission with the British Army in India. There he worked on land surveys. Back in England he taught at Sandhurst—Great Britain's West Point. He did distinguished work in clock design, weights and measures, and many other areas of instrumentation. For those accomplishments he was made a member of the Royal Society.

Kater is a creature of the eighteenth century looking upon the wonder of nineteenth-century atoms with the awe of an outsider. He published this book in 1832, when he was fifty-five. That was two years before Richard Newton signed it, and three years before he died and left the century to people such as Cooper (and, as it turns out, Richard Newton as well.)

Let us meet this Richard Newton, who entered a world where atoms would no more have to be justified by moral certainty than would Isaac Newton's laws of motion. History has not forgotten Newton. He was born in England in 1812, and his family moved to America when he was twelve. He received some manual training, but then went to college at the University of Pennsylvania. After he graduated, he became an Episcopal clergyman. He wrote voluminously. His eighteen volumes of published sermons for children were translated into a score of languages. He was a low-church evangelical, yet he helped to hold the church together when other low-church forces tried to divide it. And he was a potent opponent of slavery.

We might be tempted to confuse Richard Newton with the earlier and better-known John Newton. John was no kin but was certainly a kindred spirit, similar in many ways to Richard. John Newton was born in 1725, went to sea at eleven, and was captain of a slave ship by the age of twenty-three. When his ship almost sank during a terrible thunderstorm, he underwent a religious conversion and repudiated the slave trade. John Newton wrote the text of the hymn "Amazing Grace" and many other familiar hymn texts. He also became an Anglican clergyman and a formidable enemy of the slave trade.

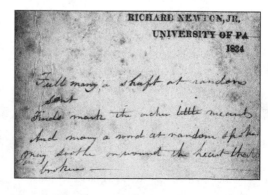

Richard Newton displayed an intensity to match that of John Newton, but he was a reconciler before he was a warrior. And that is why I have to catch my breath when I spot a bit of marginalia in this old mechanics text. He's penciled an odd poem under his name:

Full many a shaft at random sent
Finds mark the archer little meant
And many a word at random spoken
May soothe or wound the heart that's broken.

The first two lines were common currency in his time. Where the last two come from, I do not know. Whether he made them up or borrowed them, they are certainly emblematic of the life that he led after he left school. Before Newton studied theology, he studied mechanics, underlining passages about the moral force of the new atomic science and doing lever and beam-loading calculations on the flyleaves. Then he turned his attention to healing the many wounds of a world undergoing radical change.

Newton also named one of his sons William Wilberforce Newton, after the British member of the House of Commons who had been instrumental is stopping England's slave trade. Thus, once again, we glimpse, through an old technical textbook, the first halting steps that would define a life devoted to social reform.

The next logical step in spreading literacy, and of providing education as Cooper had conceived it, was the idea of correspondence school. Once more, I offer an old book as a port of entry. It's a 1904 handbook, *Mechanics' Pocket Memoranda*—a tiny book, only 3½ by 5 inches.[10] It really does fit in one's pocket. In front we find the usual stuff of the old handbooks: trig functions, logarithms, conversion tables, and such. But deeper into the book we find a set of mini-engineering textbooks with wonderful drawings.

The book was published by the International Correspondence Schools (better known as

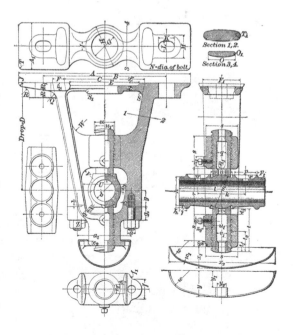

From *Mechanics' Pocket Memoranda.*

ICS), and it brings back a world that I entered in 1946. I was doing badly in high school when a successful land surveyor took me under his wing and lent me his old ICS course on drafting. I started doing the drawing assignments and handing them in to him.

A world opened up, for now no instructor was talking at me and obstructing my view of the subject. I was on my own, and education made sense. This old book brings back the sunburst of light I felt back then—freedom to learn as rapidly as I was capable of learning, coupled with the special beauty of machine drawings.

People sometimes begin histories of distance learning with Paul's epistles to the churches he founded. People have tried to teach with written messages for millennia. But America began an intense concentration on mail-order schooling near the end of the nineteenth century. Anna Ticknor, daughter of a Harvard professor, established the Society to Encourage Studies at Home in 1873.[11] During the twenty-four years that she ran the society, up to her death in 1897, it served roughly 10,000 people.

Right on her heels, a number of large universities—as well as the Chautauqua movement—took up the idea. The University of California was a latecomer in 1910, but it soon had the largest academic correspondence program in the country. One variation of the theme was the formation of a system of Working-Girls' Clubs. The 1894 *Scribner's Magazine* provides a glowing description of these new organizations.[12] They were funded in part by affordable room and board payments by young ladies working in low-salaried jobs, and in part by non-working, nonresident members. As well as room and board, they provided a natural support group, educational classes, and lectures on such topics as suffrage and the economics of women entering the workforce. Of course, they provided in-house libraries as well.

This image from *Scribner's,* and its caption, hint at the revolutionary intent of Working-Girls' Clubs. The caption reads: "In the Library, Progressive Club."

But the International Correspondence School emerged as one of the largest and most significant of these many experiments in reading and education for the working public. It began in 1890 as a home study course offered by the Pennsylvania Colliery Engineer School of Mines. It proved to be so effective, however, that it expanded and spread across America. Its Scranton headquarters was a four-story castle complete with turrets. By 1900, its huge in-house printing press was producing four tons of course notes each week. ICS could already boast a quarter million students.

A picture at the end of the handbook shows a workman studying in his cramped boardinghouse room. And I know this is no advertising gimmick, for I lived in all too many of those rooms as I too did my apprenticeship for a life in engineering. I know why that man is focused so intently upon his notes.

Today our attention swings back once more to distance learning—now to be delivered by the Internet instead of the postman. This new form is still rapidly evolving, and perhaps Internet education will one day become all that the old correspondence courses once were.

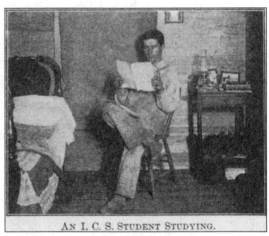

AN I. C. S. STUDENT STUDYING.

Left: The 1898 ICS home office.

Above: A student pores over an ICS lesson in the sort of cheap room typical of the times.

Another old book, a manual on how to write correspondence courses, reveals the mentality of the people who were composing such courses.[13] It says: "Questions should require the student to think, not simply recollect or repeat." Then it offers an example of a good question: "Why is a kick delivered to a dog a stimulus and one delivered to a football not a stimulus?" Well, as I mull that one, I am distracted. All I can think of is the delicious life-saving kick that this independent form of education once delivered to me.

I began by noting how surprise is common to both learning and inventing. We now uncover another theme common to both: independence. One cannot be both an inventor and a slave to the thinking of others. Invention begins and ends with rebellion: "I have a better idea; I can do it another way; I can create what has never previously existed."

At the turn of the twentieth century, the dimension of independence was still a large and explicit theme in the belief that universal learning was the way to the health of our nation. The first half of the twentieth century in America was marked by one of the great creative upsurges that the world has known.[14]

We would see one more radical and daring move in the arena of education in the twentieth century; like the Morrill Act, I have a very personal stake in this one as well. For I went to college in 1947, a time when the students around me were, on the average, ten years older than I was, for most were returning World War II veterans.

Three years before, Franklin Roosevelt had signed Public Law 346, the Servicemen's Readjustment Act—better known as the GI Bill. It gave ex-GIs tuition, books, and living expenses for college. The bill had been met with a predictable outcry against federal spending, with opponents claiming that it would provide shelter for a few slackers who did not want to go back to work. However, the government expected only 10 percent of the returning veterans to take advantage of the bill, and regarded it as relatively small potatoes, so it passed. Edwin Kiester tells about a premature headline in the *Saturday Evening Post* in 1945: "GI's Reject Education."[15] A year later one million ex-soldiers were back in school, and they kept coming. I graduated from Oregon State College in 1951, the year after the college had graduated its largest engineering class ever, by a huge margin.

Universities and other schools everywhere groaned under the load. Oregon State threw up tacky prefab quarters in their mud flats—squalid housing for married students. No rah-rah Joe College days with those people! If America expected a few slackers, they got an army of the hardest-working people I ever had the privilege of knowing. Colleges tried to cope with their numbers by running the workload up to the sky in hopes of failing enough to make room for the rest.

But those students had seen war, and compared with slaughter, this hardship was a piece of cake. I was surrounded by officers and infantrymen, back from the valley of the shadow of death. One classmate, in his late forties, had been a general. Their faces were sad, their emotions in check. And they honored me with simple academic equality.

Art Winship had suffered polio in the service. He rode his wheelchair back and forth to classes. Handicap access was unheard of in those days. The four of us nearest him grabbed his chair each morning and lugged him up two flights, for Art's disability was one more shared cost of a terrible war. Those people understood community.

The nation thus educated the survivors of war. People for whom college would have been an unimaginable privilege before the war were now in school. As the youngest member of that class, I missed out on many rites of youth, but I also knew that I was watching history unfold.

Never has a government made a wiser investment. In one stroke we further democratized education, gave new seriousness of purpose to our universities, and brought a generation back into the American mainstream. By 1956, I was back from my own turn in the army, doing graduate studies on the GI Bill and trying to sew shut the rent caused by that two-year detour. What America had done for a whole generation a few years before, she now did for me as well.

But the GI Bill was still a direct descendant of the invention of fast presses at the turn of the nineteenth century. Those presses had driven a succession of revolutionary reinventions of education—not just one revolution, not just one reinvention. We forget that at our peril as we try to ride the Internet revolution, which already has proved to be a succession of changes. The machine has changed us, and we have had to respond by changing the shape of education.

The cycle thus repeats, and out of that still-embryonic series of revolutions, I thoroughly expect a new epoch of invention—a revolution whose texture we cannot possibly begin to predict. For, if we return at last to the title of this book, let us imagine Jane Marcet dealing with it:

Mrs. B: Have we learned how invention begins?

Emily: Yes, I have an idea as to how. Does it not begin here as we converse? I was surprised that four different surfaces liberate more or less free caloric. But now I think I see a way to reduce the loss of caloric from the vessel in which I mean, one day, to travel through black space, to the Moon. Can you guess what means I have imagined for doing that?

Caroline: Emily, are you not being presumptuous? There is no such thing as a vessel that will go to the Moon.

Emily (smiling like a Cheshire cat): Indeed, I must concede that what you say is true. There is yet no such vessel.

From *Scribner's Magazine*, 1895.

14

The Arc of Invention

Finding Finished Forms

A form of Heisenberg uncertainty relation—of reality altering as we observe it—seems to confound our attempted study of the texture of invention. The more closely we have looked at the process that yields a technology, the more that process has appeared to alter. We began by asking how the airplane, steam engine, or book was invented. By the time we were done with each of these technologies, we had uncovered an increasingly filigreed fabric of interwoven invention.

We generally knew at the outset that no one person invented the airplane, the steam engine, or the book—that each had to be accumulations of radical and inspired invention. What is surprising is the complexity of the pattern formed by the many threads that combine to produce any major technology.

Yet one deconstructs myths at one's peril. As we have dissected the myths of invention we were raised with, it might seem that we have downgraded canonical inventors—those that we have named, along with those we have not. Along with Gutenberg, Watt, Fulton, and the Wrights, we could easily have questioned the roles of Morse, Bell, Edison, and all the rest, had deconstruction been our aim.

The problem is that our myths of invention, like any of our myths, are based on essential truths. I suggest that we need to find another way of looking at these adopted heroes of invention. They obviously did not create, ab initio, the technology we attribute to each. But they are important not only for their great contributions but also because, technically speaking, they mark points of crisis. Before we apply such a loaded term to invention, we need to explore the more formal meaning of the word.

We have used the word *crisis* so often to describe moments of desperation that we forget it has a far more specific meaning. A crisis is any irrevocable point in the course of things—a point at which, for better or worse, a new future trend of events is set in motion. Since the term *tipping point* captures that idea so well, Malcolm Gladwell adopted it in the place of *crisis* in his fine analysis of seemingly discontinuous shifts in the course of human affairs.[1]

Gladwell recognized that many alterations of the human condition finally do occur in that way—not step by step, but as a change that occurs abruptly after an accumulation of many steps. The camel seems comfortable under its load until we add just one straw too many. A heavy kid sits on one end of a teeter-totter, holding a skinny kid hostage at the other end, high above the ground. As the skinny kid's friends hand him one brick at a time, nothing happens. Finally they give him one last brick, and the heavy kid rises helplessly. The tipping point has been reached, and it is now his turn to be held hostage.

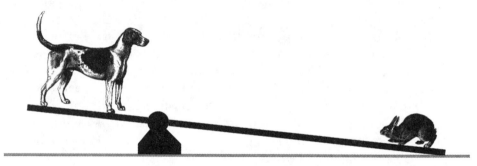

The mathematics that most of us learn in school suggests that we should expect linear behavior in the world around us—that things should be proportional to one another. Shouldn't the heavy kid rise one inch for each brick the skinny kid is handed? Well, that is not how things work.

Look at disease, says Gladwell. Imagine a twenty-four-hour flu going around. There's one chance in fifty that a sick person will transmit it when he meets a well person. If people encounter one another at the rate of forty-five people per day, the disease dies out. But at Christmastime we run into fifty-five people per day on the bus and in the malls. Things have changed only slightly, but now we face an epidemic. Consider, then, what happens if each person encounters fifty others each day. The disease now can go either way; it can die out or an epidemic can begin. Mathematically speaking, a crisis occurs at the precise point where the solution to a problem is no longer unique.

Mathematical uniqueness is another term that needs to be explained. When my generation of engineers looks back on those times when our math professors talked about the existence and uniqueness of solutions to differential equations, we remember dozing off.[2] The notion that some problems had no solution and that others might have more than one seemed fatuous. We wrote our differential equations to get answers to those problems that arose in our world. It made no sense to write equations for problems with no solution. And an equation with more than one solution seemed unthinkable, since we were in the business of describing reality with our equations. Our reality was a place where every cause begets one, and only one, effect.

We grew impatient with questions about the uniqueness or existence of solutions because we had been habituated to believe that we could rely upon Newton's laws, or the equation of heat flow, to give us one answer—at least if we posed the problem properly. Suppose, for example, that a speeding car suddenly has to negotiate a tight turn. Either it rolls over or it gets around the curve safely; it does not do both. If our math tells us it will both roll over and get around safely, then our math is silly, isn't it?

It is not. The car going around the curve actually reaches a point, a tipping point or a crisis, at which both outcomes satisfy the equations of motion. Increase the speed the tiniest bit and rolling over is certain; decrease it a hair and safety is assured. But at that one instant, the solution to the equations is not unique, and equations are powerless to say which outcome will occur.

As another example, imagine a slender glass rod, one yard long and an eighth of an inch in diameter. Stand it, on end, on the floor, then begin pressing downward on the top. Nothing happens, so we press harder. We finally reach a point where the rod suddenly bows outward and breaks. If we set up a correct mathematical solution, it will tell how much load the rod will take before it snaps. But right at that point, it can either snap or hold.

The solution is not unique in yet another way as well. Another solution has the rod bending into an S-shape instead of simply bowing outward. More solutions also exist: a triple bow, a quadruple bow, and so on. Since the fancier failure modes exist for much higher forces, the rod usually collapses by simply bowing outward suddenly and without warning. We trigger that simple failure mode when we reach the tipping point—the point of crisis.

But suppose we find a way to add a greater load all at once. Then any of those more complex collapses actually *might* occur. I once accidentally dropped such a rod on its end. When it hit the concrete

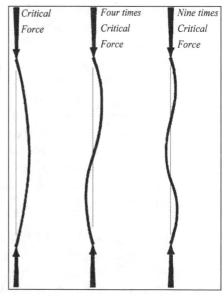

Critical Force

Four times Critical Force

Nine times Critical Force

floor below, the load was applied so suddenly that the rod collapsed in one of those wavy shapes. It broke into six pieces, each of the same length. The rod had been wavy as a snake as it collapsed, and it did so under a shock loading that was at least nine times as high as would have been needed to break the rod in simple bending.

That taught me a fine lesson about abstract math. If I might paraphrase Louis Pasteur's famous remark "Chance favors only the prepared mind," my broken glass rod helped teach me to expect surprise. Even when we think we have perfected means of prediction, outcomes still might not be unique. Nature has all kinds of unexpected alternatives up her sleeve.

Swiss mathematician Leonhard Euler solved the collapsing rod problem in the eighteenth century, when it was all the more surprising since single-valued determinism appeared to be the way of the world.[3] Now the twentieth century has given us first quantum indeterminacy and then chaos theory. We struggle to create new descriptions of a reality, which, as we begin to see, is not so straightforward. We once could yawn when math teachers began talking about existence and uniqueness. Now our ears prick up, for it has become much clearer that we no longer live in a simple world where we can imagine only one outcome.

So it is with the invention of the great technologies. The creative basis accumulates during what I call, in Chapter 10, the gestation of any major technology. That leads to a crisis, in our restrictive use of that word—a tipping point at which a canonical inventor appears.

We reach a point at which one more small step will yield a radically new future. The futures consistent with the moment are not unique—they can differ significantly from one another. The future that does emerge will depend upon which particular inventor grasps the brass ring at that critical moment.

If the brilliance of the Wright brothers had turned up in the head and hands of someone else, it might have produced a wildly different flying machine. The history of flight might have looked radically different, but the airplane was about to be invented, one way or another. We had reached the tipping point of accrued invention just as surely as an Euler column will suddenly snap into two pieces, or six, when the load is right.

In that sense it is less correct to say that this person or that invented the airplane than it is to say that what yielded the airplane was a sufficient accumulation of invention. Indeed, other claimants did invent radically different airplanes, all within a very few years of one another.

What, then, was the role of the Wright brothers? In fact they were a great deal more than just a pair of lucky names pulled out of a hat. Just as Watt brought enormous energy and invention to the steam engine, the Wrights brought those qualities to their airplanes. To pretend that they invented a machine that no one or two people ever could have invented is to denigrate their real achievement. Once we view the airplane not as an invention but as the fruit of a vast accumulation of inventions, the whole picture changes.

It might help to narrow our lens and look at just one small cluster of inventions made by the Wright brothers—those that underlay the propellers that drove their first airplane. The airplane was driven forward by two eight-foot-diameter backward-facing propellers. They were made of laminated spruce and driven by a bicycle sprocket chain. A twelve-horsepower, four-cylinder homemade engine supplied power. That internal combustion engine was the first to be made of cast aluminum, and it was also original with them.

In 2003, a group of engineers set out to reconstruct the Wrights' propellers from old records and from a few remaining fragments of old propellers.[4] The odd thing is that, as we look at the first Wright airplane, we hardly see their propellers. If we *do* notice them, they seem to be no more than narrow sticks, widening slightly toward their flat tips. Yet they represent a huge departure from anything that had been made before them.

Propellers had only recently been put to use in driving ships, but the dynamics of a propeller in water are much different than in air. Water has a thousand times the density of air. For use in water, propellers also have to be shaped in such a way that the flow will not produce seriously low-pressure regions adjacent to the blades. Such zones lead to cavitation, which eats away the surface of a ship's propeller.

It was clear to everyone that propellers had to be reconceived for use in air. The Wrights decided on propellers with airfoil cross sections. They twisted the blades to make the pitch high near the hub, where they moved slowly, and lower out at their fast-moving tips. Then they tested a reduced-scale model of the propeller, measuring its rpm and thrust, and produced performance curves. They knew that each propeller would provide sixty-seven pounds of thrust at thirty miles an hour. They also knew that would be enough to get the job done.

They made provision for changing the speed of their propellers to meet different flying conditions. Their primitive engine had no throttle, and it operated at only one speed. However, between flights, they could switch out the sprockets

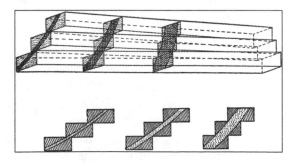

How the Wrights cut their first propellers to varying cross-sections, from three spruce boards.

to produce a higher or lower rate of rotation according to the kind of flying they anticipated—climbing or cruising, for example.

In 2003 the engineers reconstructing the propellers were able to do full-scale wind tunnel tests, which the Wrights could not do in their small tunnel. The Wrights had to rely on dimensional scaling and prediction to know how their propellers would behave in forward motion. Yet computer modeling shows that the Wrights' propeller design was almost optimal at their low airspeeds.

Old photos of propellers from the not-quite-successful airplanes built by Samuel P. Langley, Gustave Whitehead, and Hiram Maxim all show the same shape, much different from the one used by the Wrights. They used a blade shaped like a flaring triangle, and it proved to be far less successful.

Those propellers are only a fragment of a whole flying machine. Yet, by themselves, they are a microcosm of component inventions: laminations for both strength and ease of manufacturing, airfoil shaping, variable pitch, and so forth. One might be tempted to call them a component invention within the Wright brothers' airplane; but that would be incorrect since even their modest propellers were the product of many inventions.

We clearly need another word for such major accumulations of invention as the printing press, the steam engine, or the airplane. It might make sense to reserve the word *invention* for one creative act by one person, and create the word *multigenium* to describe the collective outcome of many of those individual acts—the fruit of the combined genius of many.

The process by which inventions add up to produce a multigenium is what I would like to call the *arc of inventions*. That arc progresses through a period of gestation in which inventors sense a public desire and send up various possible solutions. The cradle is the period during which the accumulation of underlying invention reaches a tipping point, and during which some key inventor arises. To that person we grant the palm, even though he or she is merely a fine exemplar—a large flash in a long arc of creative light.

The inventive hero is thus a kind of symbolic shorthand, used to get around the impossibility of naming all of the people who created the new technology—that exemplar, after which the multigenium moves fairly quickly to achieve its maturity.

This idea of a tipping point in the course of invention, by the way, carries with it a complication. Historians strongly dislike the word *inevitable*. To speak of inevitability reminds us of religious debates over predestination and free will. Yet I believe I am on solid ground in asserting that (for example) it had been inevitable since Galileo's time that gases and vapors would one day drive machines.

At the same time, the historian is correct in the sense that the Watt engine was *not* inevitable. We have seen that ideas about steam turbines and internal combustion were in people's minds long before Watt. In retrospect, we deem turbines and internal combustion engines more complex than piston steam engines, and use that complexity to explain why they were not first.

Simple primitive forms of either of those alternatives might have been made to work, however, and either one might have emerged first. It was inevitable that we would have found a way to make gases do our labor, but the exact multigenium that emerged was wide open until specific people gave it its form.

I put the question of inevitability to a historian in the following way: I asked, "Was World War I inevitable in 1912?" She replied, "War was inevitable in 1912. The particular war that we got was not." In the same way, I can flatly assert that we had reached a tipping point in 1903, and that the airplane *would be* invented, very soon.

However, Orville Wright had a near-fatal bout with typhoid fever seven years earlier. If it had killed him, a different airplane would have succeeded. Wilbur Wright might have gone ahead and made it. Maybe, instead, Santos Dumont or Glenn Curtiss would have done so. The airplane's time had come; the multigenium was ripe and ready to fall. Others knew that as well as the Wrights did, and others had comparable determination and creative genius.

At the turn of the twentieth century we had reached the tipping point at which one great inventor or another would have picked that golden apple. We would no longer be denied flight. Indeed, that sense of destiny wreathes the lives of all the canonical inventors of the various multigenia—inventors who, though never first, were pivotal.

So I say grant Morse his telegraph. Grant Prometheus his fire. Give Lindbergh his flight, and let Yogi Berra have the crown for the poetic deconstruction of English. No matter that none was first. James Watt stepped onto a well-worn road, and he brought his own special majesty to it. So did Fulton (or should I say the noncanonical John Fitch). Each did what he did very well indeed, and did so at just the right moment in history.

The long arc of inventions begins promising the birth of a major technology from the moment when possibility and desire are wed. But the arc's multicolored path can cross centuries before it reaches its pot of gold. After it has, we look back and pick our exemplar from the apogee, not from the start. Nor do we choose some maker of a finished form. We pick our hero from the high tide, where the technology first gives the appearance of functioning effectively.

What else can we do? We simply have too many names to remember—your name, my name, and then there's Ötzi. And I cannot for the life of me remember who that was, just down the hall, who came up with the marvelous idea that you and I put to use last week.

Notice, too, that I am painting the appearance of multigenia as conclusions rather than as beginnings. Our discussions of exponential growth in Chapter 8 deal with the unfolding of invention, largely after the multigenium has come into being. Invention does not stop at the apogee of the arc. Rather, it intensifies exponentially. A far greater investment of creative energy in flight came after the Wright brothers than in the years leading up to them. The multigenium is a trigger that unleashes a flood of subsequent invention.

My exponential curves are deceptive in that they are built on seeming solid data points. Each of those points would better be viewed as a tangent point where some component technology reaches the envelope formed by any such tangent curves. Each component curve represents the arc of some thread of contributing technology. In the case of steam engine efficiencies, these might be Newcomen's atmospheric engine, Watt's external-condenser-aided engines, various compound engines, and so forth. Each of those mini-arcs rose in turn to kiss the tangent line, which is built upon countless steps of human creativity.

Any one of those exponential envelopes forms a truer picture of invention than does the multigenium that marks its beginning. The way the envelope curve is shaped by the many component curves reflects the symbiosis between the many inventors and the public. It reflects the texture of the subsequent evolution of large technologies as they spin out, altering and shaping our lives.

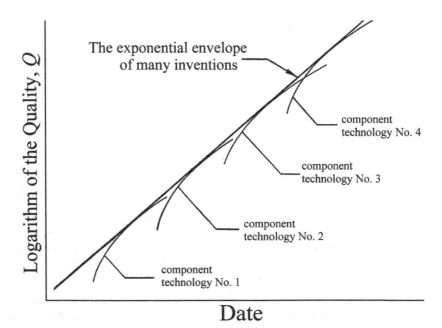

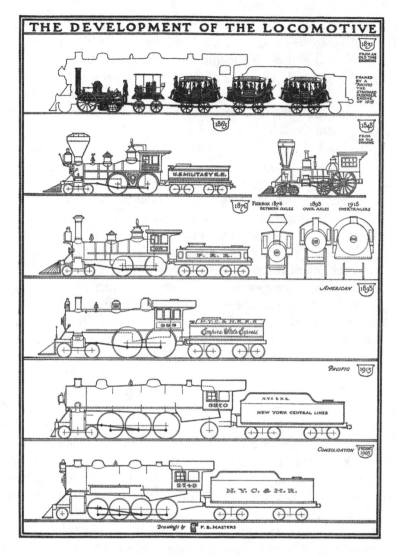

Locomotive evolution as portrayed in a 1915 book for young boys.

Once we take the view that inventions are the atoms of a multigenium—
the minimal essential building blocks, what the Greeks called the *stoichea*—
something odd happens. Just as atoms can combine to make molecules,
cells, human beings, or even galaxies, inventions can combine to make
Gutenberg's special type molds, the printed book, or modern observational science. Where does one rightly stop?

What we do is pick a technology of manageable size in the same fashion as we settle upon the canonical inventor to be the fitting representative of thousands. Take the technology of railways, for example. One might

imagine a progression of scale from components such as the engine-to-wheel linkage to the locomotive itself and then to the railway system.

We look at that progression and select the largest piece of it that we can comfortably hold in our head at one time. That would be the locomotive. So we select Robert Stephenson as the canonical inventor, and we select his *Rocket* as the exemplar of pivotal success in the development of locomotives.

In this book, I have stressed the arbitrariness of such selections by looking at several larger systems that do not lend themselves at all to such selection—inventing steam, inventing thermodynamics, inventing education. I hope that, by doing so I have helped to provide insight into the problem I suggest in my title: the problem of understanding how invention begins, and of what happens in its wake.

It is important to say, at the risk of repetition, that our aim should not be to debunk priority or disavow heroes. Indeed, the closer I look at priority or at invention's heroes, the more I see them as myths whose value is that they help us to honor originality in ourselves.

I began with Ötzi, and I would like to finish with Ötzi—that nameless Everyman who not only shapes but actually defines the human species. For we are best called not *Homo sapiens* (they who are wise) but rather *Homo technologicus* (they who deal in the lore of technique).[5] John Ruskin emphasized that in these odd words: "False things may be imagined, and false things composed; but only truth can be invented."[6]

Well, of course! When we invent, we create an uncontestable reality. We are at our best, our most human, because what we produce is what it is, and it is beyond the reach of any guile.

Emerson linked the primacy of invention in human nature to another idea I have stressed, the presence of the Zeitgeist who leads invention toward this or that multigenium. Emerson said, "Certain ideas are in the air. We are all impressionable, . . . but some more than others. . . . This explains the curious contemporaneousness of inventions and discoveries. The truth is in the air, and the most impressionable brain will announce it first, but all will announce it a few minutes later."[7]

Now, if only I were impressionable enough to hear that Zeitgeist telling me what multigenium you and I are about to create. Would it not be a wonderful thing to know what hero's face will grace the statues we make to celebrate it in AD 2300?

Nihil simul inventum est et perfectum.
Nothing is invented and perfected at the same time.

(Latin proverb)

Notes

Chapter 1

1. The literature on Ötzi is very much in flux. Some sources that I have used include S. Bortenschlager and K. Oeggl, eds., *The Man in the Ice* (New York: Springer Verlag/Wein, 2000); S. Bortenschlager and K. Oeggl, eds., *The Iceman and His Natural Environment: Palaeobotanical Results* (New York: Springer, 2000); and several websites, e.g., http://www.mummytombs.com/mummylocator/featured/otzi.news.htm, http://www.primitivearcher.com/articles/otzi-solved.html, and http://www.pbs.org/wgbh/nova/transcripts/2518iceman.html.
2. B. Bilger, "Sole Survivor," *The New Yorker,* February 14–21, 2005, 152–67.
3. L. Brisman, *Romantic Origins* (Ithaca: Cornell University Press, 1978), Introduction.

Chapter 2

1. A. Chesman, ed., *The Inventive Yankee: From Rockets to Roller Skates, 200 Years of Yankee Inventors and Inventions* (Dublin, NH: Yankee Books, 1989), 84–85. For some of the doughnut legends, see: http://www.worldwar1.com/dbc/doughnut.htm; http://id.essortment.com/doughnuthistory_rgjt.htm.
2. W. Durant, *The Story of Civilization*, Vol. I: *Our Oriental Heritage* (New York: Simon and Schuster, 1954), esp. 147.
3. The patent under discussion is J. H. Lienhard and G. S. Borkar, "Quick Opening Pressure Release Device and Method," U.S. patent no. 4,154,361, May 15, 1979. The colleague whose comment triggered the invention was Roger Eichhorn. The research underlying it is described in the following papers: J. H. Lienhard, Md. Alamgir, and M. Trela, "Early Response of Hot Water to Sudden Release from High Pressure," *Journal of Heat Transfer* 100, 3 (1978): 473–79; Md. Alamgir, C. Y. Kan, and J. H.

Lienhard, "An Experimental Study of the Rapid Depressurization of Hot Water," *Journal of Heat Transfer* 102, 2 (1979): 433–38; and Md. Alamgir, C. Y. Kan, and J. H. Lienhard, "Early Response of Pressurized Hot Water in a Pipe, to a Sudden Break," EPRI Report NP-1867, Project 687-1, June 1981.

4. S. Chartrand, "Patents," *New York Times,* August 9, 2004.
5. For more on Samuel Morey, see http://kinnexions.com/smlsource/samuel.htm.
6. A. Sutcliffe, *Steam: The Untold Story of America's First Great Invention* (New York: Palgrave Macmillan, 2004).

Chapter 3

1. A. Rice, *The Mummy, or Ramses the Damned* (New York: Ballantine Books, 1989).
2. C. Cussler, *Treasure* (New York: Pocket Books, 1988).
3. J. Vernet, "'Abbas Ibn Firnas," in C. C. Gillispie, ed., *Dictionary of Scientific Biography* (New York: Charles Scribner's Sons, 1970–1980), I:5.
4. For more on Ziryab, whose real name was Abu al-Hasan 'Ali ibn Nafi', see http://www.muslimheritage.com/topics/default.cfm?ArticleID=374 and http://www.saudiaramcoworld.com/issue/200304/flight.of.the.blackbird.htm. For more on Ibn Firnas, see http://www.mrc.org.uk/flyers.htm and http://www.angelfire.com/realm/bodhisattva/flyers.html.
5. L. White Jr., *Medieval Religion and Technology* (Los Angeles: University of California Press, 1978), Chapter 4: "Eilmer of Malmesbury, An Eleventh Century Aviator."
6. P. Scott, *The Shoulders of Giants* (Reading, MA: Addison-Wesley Publishing Co., 1995); see Chapters 2 and 3.
7. M. Hall, "Two Wings and a Prayer," *Texas Monthly,* January 2003, 100–3, 130–31.
8. T. D. Crouch, *The Eagle Aloft* (Washington, DC: Smithsonian Institution Press, 1983), Chapter 11.
9. H. Penrose, *An Ancient Air: A Biography of John Stringfellow of Chard* (Washington, DC: Smithsonian Institution Press, 1989).
10. T. D. Crouch, *A Dream of Wings: Americans and the Airplane, 1875–1905* (Washington, DC: Smithsonian Institution Press, 1989 [1981]), Chapter 5.
11. The History and Heritage Committee of the American Society of Mechanical Engineers (ASME) cites a model of a Montgomery glider in the Hiller Aircraft Museum as an International Historic Mechanical Engineering Landmark (May 11, 1996). Its history, accompanied by material unearthed after Crouch, is more favorable to Montgomery.
12. F. Howard, *Wilbur and Orville: A Biography of the Wright Brothers* (New York: Ballantine Books, 1987), especially Chapter 11.
13. Crouch, *A Dream of Wings.*
14. For more on the Loughead/Lockheed history see B. Siuru and A. Lockheed, "Lockheed: A Legacy of Speed," *Mechanical Engineering,* May

1990, 60–64, or *Lockheed Horizons* (Lockheed Corporation), 12, 7 (June 1983): 4–11. See also articles on Flora Haines Loughead in *The National Cyclopaedia* and *The Dictionary of American Biography*.

15. F. H. Loughead, "The Old Cathedral: St. Marys in San Francisco, CA," *The Ave Maria* XLVII, 24 (December 10, 1898). Available online at http://www.marysyellowstone.com/hope/The%20Old%20Cathedral.htm.

16. For more on Quimby and early women fliers, see J. H. Lienhard, *Inventing Modern: Growing Up with X-Rays, Skyscrapers, and Tailfins* (New York: Oxford University Press, 2003), Chapter 10.

17. This idea is fully developed in Lienhard, *Inventing Modern*.

Chapter 4

1. A cubic yard of the air around you weighs over two pounds.

2. I am grateful to Stein Kuiper, a lead scientist at the Phillips Research Labs in Eindhoven, the Netherlands, for pointing out the dominant role of air entrainment in yielding a cool jet. The dependence of the jet velocity on pressure drop, and the entrainment process, are both too complex to describe here. For the lesser effect—the isentropic cooling—we can write $p_1/p_2 = (T_1/T_2)^{(cp/cv - 1)/cp/cv}$. The pressures p_1 and p_2 are those in the supply reservoir (1) and the region to which the gas expands (2). T_1 and T_2 are the related absolute temperatures expressed in kelvin or degrees Rankine. The terms c_p and c_v are the specific heats for constant pressure and constant volume in the gas. For air their ratio is 1.405. Such a process is called *isentropic* because the entropy of the gas remains constant throughout the expansion.

3. T. D. Crouch, "Why Wilbur and Orville? Some Thoughts on the Wright Brothers and the Process of Invention," in R. J. Weber and D. N. Perkins, eds., *Inventive Minds* (New York: Oxford University Press, 1992), 80–92.

4. A. P. D. Mourelatos, "Empedocles of Acragas," in C. C. Gillispie, ed., *Dictionary of Scientific Biography*, Vol. IV (New York: Charles Scribner's Sons, 1971).

5. For a short account of Lucretius's life, see the article on Lucretius in C. C. Gillispie, ed., *Dictionary of Scientific Biography*, Vol. VIII (New York: Charles Scribner's Sons, 1973). See also Lucretius, *On the Nature of Things: De Rerum Natura*, ed. and trans. A. M. Esolen (Baltimore: Johns Hopkins University Press, 1995).

6. Lucretius, *On the Nature of Things: De Rerum Natura*, 59–60.

7. The Hellenistic world is dealt with in many places. For a fine introduction, see S. B. Clough, N. G. Garsoian, and D. L. Hicks, *A History of the Western World: Ancient and Medieval*, Vol. II: *The Classical World*, Part 1: *The Hellenistic World* (Boston: D. C. Heath and Company, 1964).

8. F. Klemm, *A History of Western Technology* (Cambridge, MA: MIT Press, 1964). For more on Archimedes and this view of technology, see M. Claggett, "Archimedes," in C. C. Gillispie, ed., *Dictionary of Scientific Biography*, Vol. I (New York: Charles Scribner's Sons, 1970).

9. G. Drachmann provides articles on Ctesibios (Ktesibios), Philo of Byzantium, and Hero of Alexandria in C. C. Gillispie, ed., *Dictionary of Scientific Biography* (New York: Charles Scribner's Sons, 1970–1980). For more on these inventors, see O. Mayr, *The Origins of Feedback Control* (Cambridge, MA: MIT Press, 1969), Part One, Section III.

10. L. White Jr., *Medieval Technology and Social Change* (New York: Oxford University Press, 1966), Chapter III. See also Mayr, *The Origins of Feedback Control.*

11. For some of what little has been written on Maria, see R. P. Multhauf, *The Origins of Chemistry* (New York: Franklin Watts, 1966); F. S. Taylor, *The Alchemists* (New York: Arno Press, 1974); and R. Federmann, *The Royal Art of Alchemy,* tr. R. H. Weber (Philadelphia: Chilton, 1964).

12. White, *Medieval Technology and Social Change,* Chapter 2.

13. B. Rochot, "Gassendi (Gassend), Pierre," in C. C. Gillispie, ed., *Dictionary of Scientific Biography,* Vol. V (New York: Charles Scribner's Sons, 1972).

14. For more on Cyrano's science fiction, see W. von Braun and F. I. Ordway III, *History of Rocketry and Space Travel,* rev. ed. (New York: Thomas Y. Cromwell, 1969), or read Cyrano himself in Cyrano de Bergerac, *Voyages to the Moon and the Sun,* tr. Richard Aldington (New York: Orion Press, 1962). See also Cyrano's *The Other World or the States and Empires of the Moon,* online at http://www.bewilderingstories.com/special/index.html.

15. The entries are P. L. Rose, "Magiotti, Raffaello"; M. Fichman, "Magnenus, Johann Chrysostom"; and P. L. Rose, "Magni, Valeriano," all in C. C. Gillispie, ed., *Dictionary of Scientific Biography,* Vol. IX (New York: Charles Scribner's Sons, 1974).

16. It is normal to mention Della Porta's and de Caus's conceptions in any history of the steam engine. See, e.g., H. W. Dickinson, *A Short History of the Steam Engine* (Cambridge: Cambridge University Press, 1939), or R. H. Thurston, *A History of the Growth of the Steam Engine* (Port Washington, NY: Kennikat Press, 1939 [1907]).

17. F. Krafft, "Guericke (Gericke), Otto Von," in C. C. Gillispie, ed., *Dictionary of Scientific Biography,* Vol. V (New York: Charles Scribner's Sons, 1972).

Chapter 5

1. P. P. MacLachlan, "Papin, Denis," in C. C. Gillispie, ed., *Dictionary of Scientific Biography,* Vol. X (New York: Charles Scribner's Sons, 1974).

2. Boyle's law says that the product of pressure and volume in a fixed mass of gas is constant if the gas is held at a constant temperature.

3. S. Shapin, "Who Was Robert Hooke?" in M. Hunter and S. Schaffer, eds., *Robert Hooke: New Studies* (Wolfeboro, NH: Boydell Press, 1989).

4. Ibid.

5. A nice account of Huygens's and Papin's work on the steam engine is in F. Klemm, *A History of Western Technology,* tr. Dorothea Waley Singer (Cambridge, MA: MIT Press, 1964), Part Four, "The Baroque Period," "The Struggle for a New Prime Mover."

6. Ibid. See also D. Papin, *Fasciculus dissertationum de novis quibusdam machinis atque aliis argumentis philosophicis quorum seriem versa pagina exhibet* (Marburg: J. J. Kürsnerii, 1695).

7. D. Papin, *Ars nova ad aquam ignis adminiculo efficacissime elevandam* (Frankfort-on-Main, 1707).

8. C. A. Ronan, *The Shorter Science and Civilisation in China: An Abridgement of Joseph Needham's Original Text* (Cambridge: Cambridge University Press, 1994).

9. L. Branca, *Le Machine* (1629). The Max Planck Institute has graciously put Branca's full text with images in the public domain. See http://echo.mpiwg-berlin.mpg.de/content/historymechanics/archimdesecho.

10. W. Wake, *An Exposition of the Doctrine of the Church of England in the Several Articles . . .* (London: Printed for Richard Chiswell at the Rose and Crown in St. Paul's Church-yard, 1686). For more on Wake, see J. H. Lupton, "Wake, William," in L. Stephen and S. Lee, eds., *The Dictionary of National Biography* (London: Oxford University Press, 1949–1950).

11. J. Bronowski, *The Ascent of Man* (Boston: Little, Brown and Company, 1973), Chapter 8, "The Drive for Power."

12. J. Gimpel, *The Medieval Machine* (New York: Penguin Books, 1976).

13. For excellent treatments of perpetual motion machines, see A. W. J. G. Ord-Hume, *Perpetual Motion: The History of an Obsession* (London: George Allen & Unwin Ltd., 1977); L. White Jr., *Medieval Technology and Social Change* (New York: Oxford University Press, 1966), Chapter 3.

14. A thermodynamic point here: We speak of perpetual motion machines of the first kind, PMM-I, and the second kind, PMM-II. A PMM-I violates the first law of thermodynamics by continuously producing more energy as useful work than it receives in other forms. A PMM-II need not violate the *first* law, but its efficiency is greater than allowed by the *second* law of thermodynamics.

15. See White, *Medieval Technology*, Chapter 3.

16. P. Costabel, *Leibniz and Dynamics* (Ithaca, NY: Cornell University Press, 1973).

17. H. W. Dickinson, *A Short History of the Steam Engine* (Cambridge: Cambridge University Press, 1939), Chapter 2.

18. For more on the matter of Worcester, see L. T. C. Rolt, *Thomas Newcomen: The Prehistory of the Steam Engine* (Plymouth: Latimer Trend, 1939), Chapter 1.

19. See, e.g., Dickinson, *A Short History of the Steam Engine*.

20. A very good account of Savery's work is in Rolt, *Thomas Newcomen*, Chapter 1.

21. Ibid.

22. One horsepower is the power that will lift a weight through a vertical distance at the rate of 778 foot-pounds per second.

23. The many good biographies of James Watt are generally consistent; I recommend J. P. Muirhead, *The Life of James Watt, with Selections from His Correspondence*, 2nd ed., rev. (London: John Murray, 1859); F. Arago,

Life of James Watt, 2nd ed. (Edinburgh: Adam & Charles Black, 1839); R. H. Thurston, *A History of the Growth of the Steam Engine* (New York: D. Appleton, 1888); A. Carnegie, *James Watt* (New York: Doubleday, Doran, 1933/1905), and I. B. Hart, *James Watt and the History of Steam Power* (New York: Henry Schuman, 1949).

24. See Muirhead, *The Life of James Watt.*
25. J. Boswell, *Life of Samuel Johnson,* Vol. II: *Life (1765–1776),* ed. G. B. Hill (New York: Harper and Brothers, 1891), 526.

Chapter 6

1. Since thermodynamics was a creature of the steam engine, excellent treatments of it appear in engineering textbooks. I recommend two classics: J. H. Keenan, *Thermodynamics* (New York: John Wiley and Sons, 1941 and further editions) and W. C. Reynolds and H. C. Perkins, *Engineering Thermodynamics* (New York: McGraw-Hill, 1977 and further editions). Two fine physics texts move the subject away from its steam engine kinship and say more about the connection with atomic behavior. One of the first to take account of the new quantum physics was P. Epstein, *Textbook of Thermodynamics* (New York: John Wiley and Sons, 1937). A later book offers a lucid original axiomatic formulation: H. B. Callen, *Thermodynamics and an Introduction to Thermostatistics* (New York: McGraw-Hill, 1985 and other editions). For an engineering text that links thermodynamics with atomic structure, see C. L. Tien and J. H. Lienhard, *Statistical Thermodynamics* (New York: Taylor and Francis, 1979 and other editions). And one should not overlook the various *Encyclopaedia Britannica* articles on thermodynamics.
2. A. G. Debus, "Becher, Johann Hoachim," in C. C. Gillispie, ed., *Dictionary of Scientific Biography,* Vol. 1 (New York: Charles Scribner's Sons, 1970).
3. The rare earths are Group IIIB of the periodic table. Among them is ytterbium, one of the elements used in making ceramic superconductors.
4. L. S. King, "Stahl, Georg Ernst," in C. C. Gillispie, ed., *Dictionary of Scientific Biography,* Vol. 12 (New York: Charles Scribner's Sons, 1975).
5. H. Gerlac, "Black, Joseph," in C. C. Gillispie, ed., *Dictionary of Scientific Biography,* Vol. 2 (New York: Charles Scribner's Sons, 1970).
6. H. Moore (revised by C. E. J. Herrick), "Cleghorn, George," *Oxford Dictionary of National Biography,* Vol. 12 (New York: Oxford University Press, 2004).
7. W. Cleghorn, *Disputatio physica inauguralis, theoriam ignis complectens* (Inaugural physical argument, a comprehensive theory of fire), given at Edinburgh, 1779.
8. J. Black, *Lectures on the Elements of Chemistry,* ed. J. Robison (London: Mundell and Son, 1803), esp. I:33–34. For additional background, see A. L. Donovan, *Philosophical Chemistry in the Scottish Enlightenment: The Doctrines and Discoveries of William Cullen and Joseph Black* (Edinburgh: University of Edinburgh Press, 1975).

9. For an excellent short discussion of the discovery of oxygen, see T. Kuhn, *The Structure of Scientific Revolutions* (Chicago: University of Chicago Press, 1970 [1962]), Chapter VI.

10. Epstein, *Textbook of Thermodynamics*, Chapter 2, Section 11.

11. E. Hintsche, "Haller, (Victor) Albrecht von," in C. C. Gillispie, ed., *Dictionary of Scientific Biography*, Vol. 6 (New York: Charles Scribner's Sons, 1972).

12. A. von Haller, *Versuch Schweizerischer Gedichte* (Bern: Herbert Lang, 1969). This is a facsimile of the ninth edition of Haller's poetry, published in 1762.

13. A. Haller, *First Lines of Physiology* (New York: Johnson Reprint, 1966). This is a reprint of the 1786 English edition, with an introduction by L. S. King.

14. L. S. King's introduction to Haller, *First Lines of Physiology*, xxii.

15. Haller, *First Lines of Physiology*, Section CLXXXI.

16. Ibid., Section CCLXXVII.

17. E. T. Carlson, "Rush, Benjamin," in C. C. Gillispie, ed., *Dictionary of Scientific Biography*, Vol. 9 (New York: Charles Scribner's Sons, 1974).

18. B. Rush, *Medical Inquiries and Observations Upon the Diseases of the Mind* (Philadelphia, 1812).

19. S. C. Brown, *Benjamin Thompson, Count Rumford* (Cambridge, MA: MIT Press, 1981).

20. Ibid., 198.

21. Ibid., 268.

22. C. C. Gillispie, *Lazare Carnot Savant* (Princeton, NJ: Princeton University Press, 1971), or see C. C. Gillispie, "Carnot, Lazare Nicolas-Marguerite," in C. C. Gillispie, ed., *Dictionary of Scientific Biography*, Vol. 3 (New York: Charles Scribner's Sons, 1971).

23. S. Carnot, *Reflections on the Motive Power of Fire*, ed. E. Mendoza (Gloucester, MA: Peter Smith, 1977), xii.

24. For material on Sadi Carnot, see J. F. Challey, "Carnot, Nicolas Léonard Sadi," in C. C. Gillispie, ed., *Dictionary of Scientific Biography*, Vol. 3 (New York: Charles Scribner's Sons, 1971).

25. S. Carnot, *Réflexions sur la puissance motrice du feu (Reflections on the Motive Power of Heat)*, ed. R. H. Thurston (New York: John Wiley, 1897). See also note 23.

26. An ideal gas is one for which a given amount obeys the rule $pV/T =$ constant, where p is the gas pressure, V is the volume it occupies, and T is its temperature. The air around us is very nearly ideal.

27. R. B. Lindsay, *Julius Robert Mayer: Prophet of Energy* (New York: Pergamon Press, 1973). See also R. S. Turner, "Mayer, Julius Robert," in C. C. Gillispie, ed., *Dictionary of Scientific Biography*, Vol. 9 (New York: Charles Scribner's Sons, 1974).

28. See D. S. Cardwell, *James Joule: A Biography* (Manchester: Manchester University Press, 1989); L. Rosenfeld, "Joule, James Prescott," in C. C. Gillispie, ed., *Dictionary of Scientific Biography*, Vol. 13 (New York: Charles Scribner's Sons, 1973).

29. E. E. Gaub, "Clausius, Rudolf," in C. C. Gillispie, ed., *Dictionary of Scientific Biography*, Vol. 3 (New York: Charles Scribner's Sons, 1971).

30. J. Tyndall, *Heat: A Mode of Motion*, 6th ed. (London: Longmans, Green, and Co., 1880/1863).

31. Ibid., 571.

32. Ibid., 570–71.

33. Ibid., viii.

34. Ibid., 536.

35. From R. M. Rilke, *Poems from the Book of Hours,* tr. B. Deutsch (New York: New Directions, 1941).

Chapter 7

1. A. Cazin, *La Chaleur* (Paris: Librairie Hachette, 1877 [1866]), Figure 1.

2. A. R. Hall, "'sGravesande, Willem Jacob," in C. C. Gillispie, ed., *Dictionary of Scientific Biography*, Vol. 5 (New York: Charles Scribner's Sons, 1972).

3. W. von Braun and F. I. Ordway III, *History of Rocketry and Space Travel,* rev. ed. (New York: Thomas Y. Cromwell, 1969).

4. F. H. Winter, *The First Golden Age of Rocketry* (Washington, DC: Smithsonian Institution Press, 1990).

5. B. Taylor, "Bedouin Song," *The Poetical Works of Bayard Taylor: Household Edition with Illustrations* (Boston and New York: Houghton Mifflin, 1907).

6. R. W. Emerson, "Prudence," *Essays: First and Second Series* (Boston: Houghton Mifflin, 1883).

7. P. Hone, *The Diary of Philip Hone, 1828–1851,* ed. Allan Nevins (New York: Dodd, Mead, 1936 [1927]).

8. See, e.g., T. I. Williams, *The History of Invention: From Stone Axes to Silicon Chips* (New York: Facts on File Publications, 1987), 174–75.

9. See, e.g., ibid., 174. See also *Encyclopaedia Britannica* entries.

10. J. T. Flexner, *Steamboats Come True: American Inventors in Action* (Boston: Little, Brown, 1978 [1944]), 16, 20 facing.

11. The work of Jouffroy and his antecedents is discussed in ibid., Chapter 3. See also *L'Expérience de Jouffroy d'Abbans en 1783 et la navigation à vapeur dans la région lyonnaise* (Lyon: Musée Historique de Lyon, 1983). This is a catalog of Jouffroy materials from the Lyon Historical Museum. See also M. Mollat, *Les Origines de la navigation à vapeur* (Paris: Presses Universitaires de France, 1970).

12. A. Sutcliffe, *Steam: The Untold Story of America's First Great Invention* (New York: Palgrave Macmillan, 2004).

13. Flexner, *Steamboats Come True,* 277–79.

14. Ibid., 281.

15. Sutcliffe, *Steam,* 172.

16. See Sutcliffe, *Steam,* for background and images.

17. For a nice account of transatlantic steamboat/ship crossings, see C. Wohleber, "The Annihilation of Time and Space," *American Heritage of Invention and Technology* 7, 1 (1991): 20–26. See also S. Fox, *Transat-*

lantic: Samuel Cunard, Isambard Brunel, and the Great Atlantic Steamships (New York: HarperCollins, 2003). Fox deals largely with later Atlantic crossings and stresses *Sirius*'s role as an early publicity stunt.

18. See any engineering thermodynamics textbook for details. Two are recommended in note 1 to Chapter 6.

19. C. Pursell Jr., *Early Stationary Steam Engines in America: A Study in the Migration of a Technology* (Washington, DC: Smithsonian Institution Press, 1969), 44–50.

20. S. Smiles, *The Story of the Life of George Stephenson, Railway Engineer* (London: John Murray, 1859). This is a biography of Robert Stephenson's father, George, also a noted engineer of railway systems.

21. M. R. Bailey and J. P. Glithero, *The Stephensons' Rocket: A History of a Pioneering Locomotive* (London and York: National Railway Museum and Science Museum, 2002).

22. A. Burton, *Richard Trevithick: Giant of Steam* (London: Aurum Press, 2000).

23. C. McGowan, *Rail, Steam, and Speed: The* Rocket *and the Birth of Steam Locomotion* (New York: Columbia University Press, 2004).

24. N. Faith, *The World the Railways Made* (New York: Carroll & Graf, 1990), 33–34.

25. Ibid.

26. Charles Singer, E. H. Holmyard, A. R. Hall, and Trevor I. Williams, eds., *A History of Technology* (New York: Oxford University Press, 1958), V:138–40.

27. Ibid., V:528–35.

28. F. A. Lyman, "A Practical Hero: Or How an Obscure New York Mechanic Got a Steam-Powered Toy to Drive Sawmills," *Mechanical Engineering,* February 2004, 36–38.

29. J. Bowditch, "Driving the Dynamos," *Mechanical Engineering*, April 1989, 80–89.

30. See, e.g., W. Whitman, *Leaves of Grass* (New York: Doubleday, Doran, 1940), 286.

31. See http://germansteam.info/tonup.html.

Chapter 8

1. This table was first developed in J. H. Lienhard, "The Rate of Technological Growth Before and After the 1830's," *Technology and Culture* 20, 3 (1979): 515–30. Since then, a great deal more has become known about such data. Hence many of the values here have been altered. Sources include J. Pudney, *Brunel and His World* (London: Thames and Hudson, 1974), 56, 58; http://xroads.virginia.edu/~CLASS/am485_98/sarratt/barney.html; http://www.teamvesco.com/history.htm; http://www.speedrecordclub.com/records/outright.htm. All data not included in these sources are from Chapter 7.

2. These ideas were first developed in Lienhard, "The Rate of Technological Growth," and then greatly extended in J. H. Lienhard, "Some Ideas About

Growth and Quality in Technology," *Technology Forecasting and Social Change* 27 (1985): 265–81. The latter paper is online at http://www. uh.edu/engines/qualitytechnology.pdf.

3. For details, see Lienhard, "The Rate of Technological Growth."

4. The speed records that I have given are widely available both on the Internet and in many printed sources. I have leaned particularly upon E. Angelucci, *World Encyclopedia of Civil Aircraft* (New York: Crown, 1982), the Speed Record Club website in note 1, and M. Sharpe, *Biplanes, Triplanes and Seaplanes* (New York: Barnes & Noble Books, 2000). A problem faces us here in that various groups keep their own "official" speed records. I have simply chosen the highest plausible value in each short range of dates.

5. L. T. C. Rolt, *The Aeronauts: A History of Ballooning* (New York: Walker, 1966), Chapter 10.

6. The examples in the second section of the table are from Lienhard, "The Rate of Technological Growth," and Lienhard, "Some Ideas About Growth and Quality in Technology." Those in the third section were developed by Christian L. Struble in C. L. Struble and J. H. Lienhard, "Historical Analysis of Technological Improvement: What Is Happening to Rates of Change?" University of Houston, Heat Transfer/Phase Change Laboratory, Dept. of Mechanical Engineering. 1991.

7. A more mathematical way of looking at this is to say that expectations shifted from the rate of change to its time derivative.

8. D. S. L. Cardwell, *Turning Points in Western Technology* (New York: Science History Publications, 1972), especially "Summary and Conclusions," 210–20.

9. F. L. Holmes, "Liebig, Justus von," in C. C. Gillispie, ed., *Dictionary of Scientific Biography*, Vol. 8 (New York: Charles Scribner's Sons, 1973).

10. R. M. MacLeod, "Whigs and Savants: Reflections on the Reform Movement in the Royal Society, 1830–48," in I. Inkster and J. Morrell, eds., *Metropolis and Province: Science in British Culture, 1780–1850* (Philadelphia: University of Pennsylvania Press, 1983), Chapter 2. The focus here is on how the word *professional* had been confused with political reform that was going on in the Royal Society, and it illustrates the difficulties raised by the word.

11. I develop these ideas further in "The Polytechnic Legacy," paper prepared for the ASME Management Training Workshop, Dallas, TX, August 22, 1998; available online at http://www.uh.edu/engines/asmedall.htm.

12. See, e.g., J. H. Lienhard, *Inventing Modern: Growing Up with X-rays, Skyscrapers, and Tailfins* (New York: Oxford University Press, 2003), 255–66.

Chapter 9

1. A vast amount is written about Gutenberg. My account owes much to an excellent recent book, J. Man, *Gutenberg: How One Man Remade the*

World with Words (New York: John Wiley & Sons, 2002). See also S. Füssel, *Gutenberg and the Impact of Printing*, tr. Douglas Martin (Hampshire: Ashgate, 2003).

2. For a discussion of the economics of the post-plague years in Europe, see J. Gimpel, *The Medieval Machine: The Industrial Revolution of the Middle Ages* (New York: Penguin Books, 1977).

3. D. C. McMurtrie, *The Book: The Story of Printing and Bookmaking* (New York: Oxford University Press, 1971 [1943]), 139; V. Scholderer, *Johann Gutenberg: The Inventor of Printing* (London: British Museum, 1963), 10–12.

4. McMurtrie, *The Book*, 144–47; Man, *Gutenberg*, 144–46.

5. Man, *Gutenberg*, 152–56.

6. Ibid., Plate 11. Man believes Gutenberg began large-scale printing of these indulgences in 1454.

7. A. F. West, *Alcuin and the Rise of the Christian Schools* (New York: Charles Scribner's Sons, 1909 [1892]). See also P. Wolff, *The Cultural Awakening*, tr. Anne Carter (New York: Pantheon Books, 1968), esp. Part One, "The Time of Alcuin." For an example of Alcuin's writing and his interaction with Charlemagne, see W. S. Howell, *The Rhetoric of Alcuin and Charlemagne* (New York: Russell & Russell, 1965). This is presented in the form of a dialog between Alcuin and Charlemagne.

8. Wolff, *The Cultural Awakening*, 48–53.

9. J. Marchand, "The Frankish Mother: Dhuoda," in Katharina M. Wilson, ed., *Medieval Women Writers* (Athens: University of Georgia Press, 1984), 1–29.

10. See, e.g., Wolff, *The Cultural Awakening*, Part 3, "The Time of Abelard." For a popular account of the Christian occupation of Toledo, see J. Burke, *The Day the Universe Changed* (Boston: Little, Brown, 1985), Chapter 2, "In the Light of the Above."

11. For an excellent description of how monastery scriptoria functioned, see M. Drogin, *Anathema: Medieval Scribes and the History of Book Curses* (Montclair, NJ: Allanheld & Schram, 1983).

12. J. M. Sheppard, "The Twelfth-Century Library and Scriptorium at Buildwas: Assessing the Evidence," in D. Williams, ed., *England in the Twelfth Century* (Bury St. Edmunds, Suffolk: St. Edmundsbury Press, 1990), 193–204.

13. K. W. Humphreys, *The Book Provisions of the Mediaeval Friars: 1215–1400* (Amsterdam: Erasmus Booksellers, 1964), 14–17.

14. For more on scriptoria and their transition from the monastery to the public domain, see M. A. Rouse and R. H. Rouse, *Authentic Witnesses: Approaches to Medieval Texts and Manuscripts* (Notre Dame, IN: University of Notre Dame Press, 1991), Chapter 8, "The Book Trade at the University of Paris."

15. G. Sarton, *Galen of Pergamon* (Lawrence: University of Kansas Press, 1954), Chapter 2.

16. The Archimedes palimpsest is best viewed/explained online. See, e.g., http://www.pbs.org/wgbh/nova/archimedes/palimpsest.html.
17. J. Dreyfus, "The Invention of Spectacles and the Advent of Printing," in *Into Print: Selected Writings on Printing History, Typography and Book Production* (London: British Library, 1994), 298–310.
18. J. M. Bloom, *The History and Impact of Paper in the Islamic Lands* (New Haven: Yale University Press, 2001). See also J. M. Bloom, "Revolution by the Ream: A History of Paper," *Aramco World,* May/June 1999, 26–39.
19. For additional background on paper, see, e.g., R. Temple, *The Genius of China* (New York: Simon & Schuster, 1986), 81–84; L. B. Schlosser, "A History of Paper," in Paulette Long, ed., *Paper—Art and Technology* (San Francisco: World Print Council, 1979), 2–19; T. I. Williams, *The History of Invention* (New York: Facts on File Publications, 1987), Chapter 10, "Paper and Printing"; T.-h. Tsien, *Written on Bamboo and Silk* (Chicago: University of Chicago Press, 1962).
20. Bloom, "Revolution by the Ream."
21. H. Lehmann-Haupt, *Gutenberg and the Master of the Playing Cards* (New Haven: Yale University Press, 1966).
22. See Temple, *The Genius of China,* 110–19; T. F. Carter and L. C. Goodrich, *The Invention of Printing in China, and Its Spread Westward* (New York: Ronald Press, 1955). In both sources, the authors attempt to trace links between Chinese printing with movable type and Gutenberg's invention but are unable to conclusively show one.
23. Drogin, *Anathema.* Alcuin's line went thus: "*O quam dulcis vita fuit dum sedebamus quieti . . . inter librorum copias.*" I have recast it in haiku form.

Chapter 10

1. E. Angelucci, *World Encyclopedia of Civil Aircraft* (New York: Crown, 1982).
2. See, e.g., J. Shurkin, *Engines of the Mind* (New York: W. W. Norton, 1984), Chapter 1.
3. Z. G. Pascal, *Endless Frontier: Vannevar Bush, Engineer of the American Century* (New York: Free Press, 1997).
4. The internal storage of user programs is an important landmark in the computer's cradle years, and one seldom noted in computer histories. In 1948, a prototype called Baby was developed at the University of Manchester. It was the first computer to store a user program as well as data. Details at http://www.computer50.org.
5. See, e.g., T. R. Reid, *The Chip: How Two Americans Invented the Microchip and Launched a Revolution* (New York: Simon and Schuster, 1984); M. S. Malone, *The Microprocessor: A Biography* (New York: Springer Verlag/ TELOS, 1995).
6. K. W. Humphreys, *The Book Provisions of the Mediaeval Friars: 1215– 1400* (Amsterdam: Erasmus Booksellers, 1964), 14–17.
7. M. Drogin, *Anathema: Medieval Scribes and the History of Book Curses* (Montclair, NJ: Allanheld & Schram, 1983).

8. Ibid., 66.
9. A. W. Pollard, *An Essay on Colophons, with Specimens and Translations* (New York: B. Franklin, 1968).
10. From S. Brant, "The Printer's Art," in A. Sparrow and J. S. Perosa, eds., *Renaissance Latin Verse: An Anthology* (Charlotte: University of North Carolina Press, 1979), no. 252. I am grateful to Dora Pozzi, University of Houston Classics Department, for her translation.
11. J. Moran, *Printing Presses: History and Development from the Fifteenth Century to Modern Times* (Berkeley: University of California Press, 1973), Chapter 1.
12. This is the first third of the explicit, with the words changed from Caxton's late Middle English into modern spelling. The full original is given in W. Blades, *The Life and Typography of William Caxton: England's First Printer* (London: Joseph Lilly, 1861–1863), 1:134.
13. Much is written about Caxton. In addition to Blades's nineteenth-century treatise, see, e.g., L. Hellinga, *Caxton in Focus: The Beginning of Printing in England* (London: British Library, 1982); G. E. Painter, *William Caxton: A Biography* (New York: G. P. Putman's Sons, 1977); E. Childs, *William Caxton: A Portrait in a Background* (London: Northwood Publications, 1976); N. F. Blake, *Caxton: England's First Publisher* (London: Osprey, 1976); and R. A. Deacon, *A Biography of William Caxton: The First English Editor, Printer, Merchant and Translator* (Chatham: Frederick Muller, 1976).
14. C. Ginzburg, *The Cheese and the Worms: The Cosmos of a Sixteenth-Century Miller,* tr. John and Anne Tedeschi (London and Henley: Routledge & Kegan Paul, 1980).

Chapter 11

1. L. Carroll, *Alice in Wonderland*. Taken (appropriately enough) from the Gutenberg Project posting: http://www-2.cs.cmu.edu/People/rgs/alice-table.html.
2. A. Dürer, *Underweysung der Messung, mit dem Zirckel un Richtscheyt, in Linien* (Nuremberg, 1525). For a reprint with accompanying translation see A. Dürer, *The Painter's Manual,* translated with commentary by W. L. Strauss (New York: Abaris Books, 1977). The section on letters is also available: A. Dürer, *Of the Just Shaping of Letters: From the Applied Geometry of Albrecht Dürer* (New York: Dover Publications, 1965).
3. C. L. Svensen, *Drafting for Engineers,* 2nd ed. (New York: D. van Nostrand, 1941), 75.
4. Dürer, *Of the Just Shaping of Letters,* 5.
5. Dürer, *The Painter's Manual,* 22–24.
6. Dürer, *Of the Just Shaping of Letters,* 2.
7. See, e.g., D. Hassig, *Medieval Bestiaries: Text, Image, Ideology* (New York: Cambridge University Press, 1995).
8. J. Veldener, *Herbarius* (Louvain: Peter Schoeffer, 1484).

9. L. Hellinga, *Caxton in Focus: The Beginnings of Printing in England* (London: British Library, 1982), discusses the relation between Caxton and Veldener.

10. *Herbarium* (Passau: Johann Petri, 1486).

11. H. Schedel, *Liber Chronicarum* (Nuremberg: Anton Koberger, 1493).

12. K.-A. Knappe, *Dürer: The Complete Engravings, Etchings, and Woodcuts* (Secaucus, NJ: Wellfleet Press, 1965), x.

13. E. H. Zeydel, *Sebastian Brant* (New York: Twayne Publishers, 1967).

14. For a particularly useful picture of the Italian use of perspective at the time of this visit by Dürer, see W. M. Ivins Jr., *On the Rationalization of Sight: With an Examination of Three Renaissance Texts on Perspective* (New York: Da Capo Press, 1973).

15. Euclid, *Elementa Geometriae* (Venice: Erhard Ratdolt, 1482), reproduced in *Art of the Printed Book: 1455–1955* (New York: Pierpont Morgan Library, 1973).

16. C. D. O'Malley and J. B. de C. M. Saunders, *Leonardo da Vinci on the Human Body* (New York: Henry Schuman, 1953).

17. E. Belt, *Leonardo the Anatomist* (New York: Greenwood Press, 1969 [1955]). See also S. B. Nuland, *Doctors* (New York: Vintage Books, 1989), 68–70; K. Clark, *Leonardo Da Vinci*, rev. ed. (New York: Penguin Books, 1988), 67.

18. J. P. Richter, ed., *Notebooks of Leonardo da Vinci* (New York: Dover, 1970): II:xiv, 105.

19. Belt, *Leonardo the Anatomist*, 67.

20. For more on Vesalius, see A. Vesalius, *De Humani Corporis Fabrica Libri Septem* (Basel: Johannes Oporinus, 1543); C. D. O'Malley, *Andreas Vesalius of Brussels, 1514–1564* (Berkeley: University of California Press, 1965); S. B. Nuland, *Doctors: The Biography of Medicine* (New York: Vintage Books, 1988).

21. For more on Galen, see, e.g., Galen, *Galen on Anatomical Procedures*, translated with an introduction by C. Singer (London: Oxford University Press, 1956); G. Sarton, *Galen of Pergamon* (Lawrence: University of Kansas Press, 1954); Nuland, *Doctors*.

22. A. Vesalius, *The Epitome of Andreas Vesalius*, translated and with an introduction by L. R. Lind, anatomical notes by C. W. Asling, foreword by L. Clendening (Cambridge, MA: MIT Press, 1949), xxxiii–xxxvi.

23. Paracelsus, *Selected Writings*, edited with an introduction by J. Jacobi, translated by N. Guterman (New York: Pantheon Books, 1951). See also H. M. Pachter, *Paracelsus: Magic into Science* (New York: Henry Schuman, 1951).

24. Paracelsus, *Selected Writings*, 123.

25. E. L. Eisenstein, *The Printing Press as an Agent of Change: Communications and Cultural Transformations in Early-Modern Europe* (New York: Cambridge University Press, 1979), II: Chapter 6.

26. Personal communication from medical historian Helen Valier.

27. P. G. Bietenholz, "Printing and the Reformation in Antwerp," in J.-F. Gilmont, ed., *The Reformation and the Book,* translated by K. Maag (Aldershot: Ashgate, 1998).

28. For a fine insight into the way the argument was still spinning out in the mid-seventeenth century, see C. Zimmer, *Soul Made Flesh: The Discovery of the Brain—and How It Changed the World* (New York: Free Press, 2004). Zimmer shows how a great scientific shift occurred at Oxford University, during and right after the English Civil War.

29. A. Paré, *The Apologie and Treatise of Ambroise Paré . . . ,* edited by G. Keynes (Chicago: University of Chicago Press, 1952). Originally published in 1585, this is an edited facsimile of a 1634 English edition.

30. Ibid., 23–24.

31. Paré's *Apologie and Treatise* includes a seven-page polemic against firearms, 130–36.

32. A. Paré, *On Monsters and Marvels,* translated by J. L. Pallister (Chicago: University of Chicago Press, 1982).

33. F. Bacon, *Novum Organum,* translated and edited by P. Urbach and J. Gibson (Chicago: Open Court, 1993).

34. From W. Wordsworth, "Lines Composed a Few Miles Above Tintern Abbey," *Selected Poems of William Wordsworth,* edited by H. R. Steeves (New York: Harcourt, Brace and Company, 1922), 122–26.

Chapter 12

1. J. H. Marcet, *Conversations on Chemistry; in Which the Elements of that Science Are Familiarly Explained and Illustrated by Experiment,* 4th ed, Vol. I: *On Simple Bodies* (London: Longman, Hurst, Rees, Orme, and Brown, 1813), 61–66.

2. Ibid., vii.

3. J. Moran, *Printing Presses: History and Development from the Fifteenth Century to Modern Times* (Berkeley: University of California Press, 1973), Chapters 3–5.

4. Ibid., 101–3.

5. Much of what we know Mary Anne Howley comes from mentions in articles on William Howley and Sir George Howland Beaumont in H. C. C. Matthew and B. Harrison, eds., *The Dictionary of National Biography* (New York: Oxford University Press, 2004). My thanks to University of Houston librarian Barbara Kemp for locating additional fragmentary online biographical information about Howley.

6. See the article on Charles Leslie in *The Dictionary of National Biography.*

7. T. G. West, *In the Mind's Eye* (New York: Prometheus Books, 1991), especially Chapter 4.

8. J. Tyndall, *Faraday as a Discoverer* (London: Longmans, Green, 1868), 6–7.

9. T. P. Jones, *New Conversations on Chemistry adapted to the present state of that science . . .* (Philadelphia: John Grigg, 1833), 105. Jones was an American educator who edited Marcet's book and added questions for

classroom use. He made no secret of Marcet's identity in his preface, although he does not include her name on the title page.

10. See, e.g., J. H. Marcet, *Conversations on Natural Philosophy* (Boston: Lincoln, Edmonds, 1834). This, one of many later editions of Marcet's book, bears the name of J. L. Blake on its title page and is sometimes catalogued under his name.

11. My thanks to librarian and genealogist Stephen Perkins for tracing Joseph L. Whittenburg.

12. J. L. Comstock, *A System of Natural Philosophy* (New York: Pratt, Woodford, 1848).

13. D. Hunter, *Papermaking: The History and Technique of an Ancient Craft* (New York: Alfred A. Knopf, 1943), Chapter 11. For more on papermaking, see also C. Singer, E. J. Holmyard, A. R. Hall, and T. I. Williams, *A History of Technology,* Vol. V: *The Late Nineteenth Century, c. 1850–c. 1900* (London: Oxford University Press, 1958), 712–13.

14. S. Jennett, *Pioneers in Printing* (London: Routledge & Kegan Paul, 1958), 47–58.

15. D. Lardner, *Hand-Book of Natural Philosophy* (London: Walton and Maberly, 1856). William James's copy of this book, with his marginal notations, is in the Houghton Library at Harvard University.

16. T. C. Sterling, "Marcet's Apparatus," *Rittenhouse: Journal of the American Scientific Instrument Enterprise* 8, 4 (1994): 110–13.

17. C. Davies, *Elements of Geometry and Trigonometry, from the Works of A.M. Legendre.* Revised and Adapted to the Course of Mathematical Instruction in the United States (New York: A. S. Barnes, 1855).

18. I am most grateful to Caroline Eastman of the Portsmouth Athenaeum in Portsmouth, New Hampshire, for her help in finding local biographical material on Foster.

19. For more on these ships, see: T. Gibbons, *Warships and Naval Battles of the Civil War* (New York: W. H. Smith, 1989).

Chapter 13

1. For Cooper's story, see E. C. Mack, *Peter Cooper: Citizen of New York* (New York: Duell, Sloan and Pearce, 1949), or M. Gurko, *The Lives and Times of Peter Cooper* (New York: Thomas Y. Crowell, 1959). See also Note 2 in this chapter.

2. A. Nevins, *Abram S. Hewitt, with Some Account of Peter Cooper* (New York: Harper & Brothers, 1935).

3. Ibid., 448.

4. Gurko, *The Lives and Times of Peter Cooper,* 163.

5. E. G. Hewitt, *The Making of a Modern Museum* (New York: Wednesday Afternoon Club, 1919), available online at http://www.ringwoodmanor.com/peo/ch/nellie/mmm.htm.

6. H. Kater and D. Lardner, *A Treatise on Mechanics* (Philadelphia: Carey & Lea, 1832).

7. "Newton, Richard," in D. Malone, ed., *Dictionary of American Biography* (New York: Charles Scribner's Sons, 1934), 13:474.
8. Kater and Lardner, *A Treatise on Mechanics,* 12.
9. "Kater, H.," in L. Stephen and S. Lee, eds., *Dictionary of National Biography* (London: Smith, Elder, 1908–1909), Vol. 13.
10. *Mechanics' Pocket Memoranda: A Convenient Pocketbook,* 6th ed. (Scranton, PA: Colliery Engineer Company, 1900).
11. O. MacKenzie, E. L. Christensen, and P. H. Rigby, *Correspondence Instruction in the United States* (New York: McGraw-Hill, 1968).
12. C. S. Davidge, "Working-Girls' Clubs," *Scribner's Magazine,* May 1894, 619–28.
13. G. R. Potter, *Planning and Writing Correspondence Courses* (Berkeley: University of California Press, 1946), 13–14.
14. This idea is the subject of the book: J. H. Lienhard, *Inventing Modern: Growing up with X-Rays, Skyscrapers, and Tailfins* (New York: Oxford University Press, 2003).
15. E. Kiester Jr., "The G.I. Bill May Be the Best Deal Ever made by Uncle Sam," *Smithsonian,* 1994, pp. 128–39.

Chapter 14

1. M. Gladwell, *The Tipping Point: How Little Things Can Make a Big Difference* (Boston: Back Bay Books, 2002).
2. For some discussion of mathematical existence and uniqueness, see, e.g., I. N. Sneddon, *Elements of Partial Differential Equations* (New York: McGraw-Hill, 1957), 9, 48.
3. For a clear account of Euler column instability, see A. J. S. Pippard, *The Analysis of Engineering Structures* (London: Arnold, 1957); E. P. Popov, *Introduction to Mechanics of Solids* (Englewood Cliffs, NJ: Prentice-Hall, 1869).
4. R. A. Ash, C. P. Britcher, and K. W. Hyde, "Prop-Wrights: How Two Brothers from Dayton Added a New Twist to Airplane Propulsion," in "100 Years of Flight," supplement to *Mechanical Engineering,* December 2003, 11–15, 39. See also http://www.wrightexperience.com/pdfs/props.pdf.
5. J. H. Lienhard, *The Engines of Our Ingenuity: An Engineer Looks at Technology and Culture* (New York: Oxford University Press, 2000), Chapter 1.
6. J. Ruskin, *Modern Painters* (London: G. Allen, 1898–1904), Part VIII, Chap. 4, Sec. 23.
7. R. W. Emerson, *The Conduct of Life* (New York: Houghton Mifflin, 1904 [1860]), 44.

Illustration Credits

Page 3
From H. R. Hall, *The Threshold of History* (New York: Funk & Wagnalls, 1914).

Page 10
Indian vada and New Orleans beignet. Photos by author.

Page 12
Limestone bas-relief of Imhotep (or Hotepa), from *Encyclopaedia Britannica,* 1911.

Page 13
Cross section of the Stepped Pyramid, from J. Fergusson, *A History of Architecture in All Countries* (New York: Dodd, Mead, 1883).

Page 21
Daedalus, now in the Museum of Science, Boston. Photo by author.

Page 23
Birds. Photo by author.

Page 29
Left: Otto Lilienthal aloft in the same glider in which he crashed and died in 1896.
Right: One of Octave Chanute's gliders, probably in the same year. Both pictures are from W. Kaempffert, *The New Art of Flying* (New York: Dodd, Mead and Co., 1911).

Page 32
Miller Aerostat, 1843, from H. Golding, *The Wonder Book of Aircraft for Boys and Girls* (London: Ward, Lock & Co., 1920).

Page 36
Steam leaving a locomotive cylinder. Photo by author.

Page 44
Approximate configuration of Hero's turbine. Sketch by author.

Page 45
A plain domestic double boiler, on sale in the 1900 Sears, Roebuck and Co. catalog.

Illustration Credits

Page 49
Representations of two of von Guericke's experiments (Magdeburg sphere and lifting a heavy weight by drawing a vacuum in a cylinder into which was fitted a piston) by J. D. Steele, *Fourteen Weeks in Physics* (New York: A. S. Barnes & Co., 1878).

Page 51
Sculpture of Denis Papin. Photo by author.

Page 53
Papin's sketch of his first steam engine, from *Ars nova ad aquam ignis adminiculo efficacissime elevandam* (Frankfort-on-Main, 1707).

Page 54
Papin's final steam engine design, as drawn by D. Lardner in *The Steam Engine Familiarly Explained and Illustrated* . . . (Philadelphia: E. L. Carey & A. Hart, 1836).

Page 56
Detail of engraving. From L. Branca, *Le Machine* (1629).

Page 59
Author's overcentered wheel type of perpetual motion machine.

Page 62
Savery's 1699 version of his steam pump, as shown by D. Lardner in *The Steam Engine Familiarly Explained and Illustrated* . . . (Philadelphia: E. L. Carey & A. Hart, 1836).

Page 64
Newcomen's atmospheric engine, as shown by D. Lardner in *The Steam Engine Familiarly Explained and Illustrated* . . . (Philadelphia: E. L. Carey & A. Hart, 1836).

Page 65
Later-eighteenth-century Newcomen-type piston in London Science Museum. Photo by author.

Page 65
The Glasgow model of a Newcomen engine presented to Watt for rework, from R. H. Thurston, *A History of the Growth of the Steam Engine* (New York: D. Appleton, 1888).

Page 67
Watt's first engine design, from D. Lardner, *Popular Lectures on the Steam Engine* . . . (New York: Elam Bliss, 1828).

Page 68
Left: A pantograph device from T. E. French, *A Manual of Engineering Drawing for Students and Draftsmen,* 2nd ed. (New York: McGraw-Hill, 1918). *Right*: A model of Watt's straight-line mechanism showing how rods are arranged to produce an almost perfectly vertical motion of the engine piston at point A. This model was built by Simon Dorton as an exercise in his history of technology course with the author.

Page 69
A more fully evolved Watt engine, from D. Lardner, *Popular Lectures on the Steam Engine* . . . (New York: Elam Bliss, 1828).

Page 74
Wax turning into phlogiston and soot. Photo by author.

Page 77
Two examples of thermal property observations from A. Cazin, *La Chaleur* (Paris: Librairie Hachette, 1877 [1866]), Figures 58 and 70.

Page 79
Creating caloric. Photo by author.

Illustration Credits

Page 80

From Haller's poetry, part of a sad ode to his first wife, Mariane, who died the year he moved to Göttingen. A. von Haller, *Versuch Schweizerischer Gedichte* (Bern: Herbert Lang, 1969). This is a facsimile of the ninth edition of Haller's poetry, published in 1762.

Page 87

Carnot's steam engine waterwheel analogy. Sketch by author.

Page 88

This copy of a widely circulated image of Robert Julius Mayer (1814–1878) is taken from M. Gumpert, *Das Leben fur die Idee* (Berlin: S. Fischer Verlag, 1935).

Page 89

Joule's apparatus for measuring the temperature rise of water churned by a paddle wheel in the London Science Museum. Photo by author.

Page 95

Éolipyle à recul. A. Cazin, *La Chaleur* (Paris: Librairie Hachette, 1877 [1866]): Figure 1.

Page 97

Part of a Congreve rocket fired during the Battle of Stonington, August 9–12, 1814, by the British ship *Terror* in Stonington Lighthouse Museum, Stonington, CT. Photo by author.

Pages 99 and 100

Rain, Steam, and Speed—The Great Western Railway, by J. M. W. Turner. This black-and-white reproduction of Turner's vision of speed was published in the same year the Wright brothers flew, in C. Holme, ed., *The Genius of J. M. W. Turner, R. A.* (New York: The Studio, 1903), Plate O–23.

Page 101

Hulls's steam tug as he represented it in his advertising notice, from D. Brewster, ed., *The Edinburgh Encyclopaedia* (Philadelphia: Joseph and Edward Parker, 1832), Plate DX.

Page 103

An 1832 image of Fitch's first boat, from D. Brewster, ed., *The Edinburgh Encyclopaedia* (Philadelphia: Joseph and Edward Parker, 1832), Plate DXI.

Page 105

Model showing the structure of Fulton's *North River Steamboat (Clermont)* in the London Science Museum. Photo by author.

Page 106

An oceangoing steamship in an 1832 conception—after *Savannah* but still before *Great Western*, from D. Brewster, ed., *The Edinburgh Encyclopaedia* (Philadelphia: Joseph and Edward Parker, 1832), Plate DX.

Page 109

Detail of the rack-and-pinion drive still in use in New Hampshire on the Mt. Washington cog railway today. From the Mt. Washington museum. Photo by author.

Page 110

Stephenson's *Rocket*. *Left*: As represented by S. Smiles in his 1859 biography of Stephenson's father, *The Story of the Life of George Stephenson, Railway Engineer* (London: John Murray, 1859). *Right*: A scale model in the Boston Museum of Science. Photo by author.

Page 111

Typical steam locomotive firebox on a locomotive, still used on the Silverton Durango Railway. Photo by author.

Page 113

A cutaway 1891 Parsons axial-flow turbine, revealing a series of dozens of stages of turbine blades, in the London Science Museum. Photo by author. The *Turbinia*, from R. M. Nielson, *The Steam Turbine* (New York: Longmans, Green, 1905 [1902]).

Illustration Credits

Page 114
Empire State Express No. 999 in the Chicago Museum of Science and Industry. Photo by author.

Page 142
Gutenberg Bible facsimile, Houston Museum of Printing History. Photo by author.

Page 144
Example of ninth-century Carolingian script, shown in the 1897 *Encyclopaedia Britannica*.

Page 151
Left: S. Brant, *Das Narrenschiff* (Basel: Johann Bergmann von Olpe, 1494), from a Latin version published in 1497. Special Collections, University of Houston Libraries. *Right*: A typical pair of glasses offered for sale to ordinary people in the 1895 Montgomery-Ward catalog.

Page 152
A 1/3-inch length of bond paper, torn against its grain. Scan by author.

Page 153
European papermaking in the seventeenth century. From G. A. Böckler, *Theatrum Machinarum Novum, das ist: Neuvermehrter Schauplatz der Mechanischen Kûnsten* (Nuremberg, 1673). Courtesy of Special Collections, University of Kentucky.

Pages 163–64
Replica of a late-fifteenth-century "Gutenberg press" made by Steven Pratt, Pratt Wagon Works. Museum of Printing History, Houston, Texas. Photo by author.

Page 165
Simple woodblock image of a strict tutor and four pupils, used repetitively by William Caxton in several books during the late fifteenth century. From W. Blades, *The Life and Typography of William Caxton: England's First Printer* (London: Joseph Lilly, 1861–63), vol. 2, Plate LVI.

Page 170
Hubble photograph of galaxies curdling out of the deep field some 13 billion light-years from the earth. Courtesy of NASA.

Page 176
Detail of a Veldener page, after L. Hellinga, *Caxton in Focus: The Beginnings of Printing in England* (London: British Library, 1982), 70.

Page 176
A recognizable oxtongue plant in Petri's herbal. Woodcut with watercolor. Image courtesy of Special Collections, University of Houston Libraries.

Page 177
Representation of a geranium in a leaf from a Bohemian herbal, ca. 1516. Image courtesy of Special Collections, University of Houston Libraries.

Page 178
The Arnolfini Wedding, Jan van Eyck, 1434 (from J. C. M. Weale, *Van Eyck* [London, T. C. & E. C. Jack, ca. 1912], 40), and a detail of the spherical mirror reflection (from W. H. J. Weale and M. W. Brockwell, *The Van Eycks and Their Art* [London: John Lane, 1912], Plate XXV).

Page 179
The city of Trier, as represented by Schedel in *Liber Chronicarum* (Nuremberg: Anton Koberger, 1493).

Page 181
St. Jerome in His Study, 1514, from V. Scherer, *Dürer: Des Meisters Gemälde Kupeferstiche und Holzschnitte* (Stuttgart: Deutsche Verlags-Anstalt, 1904), 130.

Illustration Credits

Page 182
St. Anthony, 1519, from V. Scherer, *Dürer: Des Meisters Gemälde Kupeferstiche und Holzschnitte* (Stuttgart: Deutsche Verlags-Anstalt, 1904), 143.

Page 182
Dürer's graphical derivations of the forms of ellipses and parabolas from a cone, from his *Underweysung der Messung, mit dem Zirckel un Richtscheyt, in Linien* (Nuremberg, 1525).

Page 186
From A. Vesalius, *De Humani Corporis Fabrica Libri Septem* (Basel: Johannes Oporinus, 1543).

Page 187
A Vesalius skeleton, from *De Humani Corporis Fabrica Libri Septem* (Basel: Johannes Oporinus, 1543).

Page 190
Instruments for suturing with wire, from A. Paré, *Three and Fifty Instruments of Chirurgery*, from the English tr. (H. C. London printed for Michael Sparke, 1631).

Page 191
A bullet puller, from A. Paré, *Three and Fifty Instruments of Chirurgery*, from the English tr. (H. C. London printed for Michael Sparke, 1631).

Page 197
Pictet's apparatus for the reflection of heat in Jane Haldimand Marcet, *Conversations on Chemistry; in Which the Elements of That Science Are Familiarly Explained and Illustrated by Experiment*, 4th ed, Vol. I: *On Simple Bodies* (London: Longman, Hurst, Rees, Orme, and Brown, 1813), Plate III.

Page 198
The first of Stanhope's all-metal presses, from D. Brewster, ed., *The Edinburgh Encyclopaedia*, Vol. III (Philadelphia: Joseph and Edward Parker, 1832).

Page 199
Clymer's Columbian press. The line drawing is from D. Brewster, ed., *The Edinburgh Encyclopaedia*, Vol. III (Philadelphia: Joseph and Edward Parker, 1832). The author's photos (taken at the Houston Museum of Printing History) show details of Clymer's eagle counterweight and his complex three-dimensional lever system.

Page 200
Above: A Koenig double roller press, manufactured by Applegath and Cowper in London before 1827. *Below*: The inking scheme, and the movement of paper, in this press. From C. Tomlinson, ed., *Cyclopaedia of Useful Arts, Mechanical and Chemical, Manufactures, Mining, and Engineering* (New York: George Virtue, 1854), 487–90.

Page 201
Conversations in Chemistry. Photo by author.

Page 204
Michael Faraday, from J. Tyndall, *Faraday as a Discoverer* (London: Longmans, Green, 1868), frontispiece.

Page 205
Marcet shows how one's eye sees a bug through a compound microscope. From J. H. Marcet, *Conversations on Natural Philosophy* (Boston: Lincoln, Edmonds, 1834), Plate XXII.

Page 207
Wasp paper. Photo by author.

Page 210
Marcet's Apparatus from J. Johnston, *A Manual of Chemistry* (Philadelphia: Charles Desilver, 1856 [1840]), 43. (This book was also done in stereotype.)

Illustration Credits

Page 217
Peter Cooper, from the 1897 *Encyclopaedia Britannica* supplement.

Page 219
Model of the *Tom Thumb* in the Museum of Science, Boston. Photo by author.

Page 220
Abram S. Hewitt, from the 1897 *Encyclopaedia Britannica* supplement.

Page 227
Mechanics' Pocket Memoranda: A Convenient Pocketbook, 6th ed. (Scranton, PA: Colliery Engineer Company, 1900), 203.

Page 228
This image from *Scribner's,* and its caption, hint at the revolutionary intent of Working-Girls' Clubs. The caption reads: "In the Library, Progressive Club." C. S. Davidge, "Working-Girls' Clubs," *Scribner's Magazine,* May 1894, 619–28.

Page 229
Left: The 1898 ICS home office. *Above:* A student pores over an ICS lesson in the sort of cheap room typical of the times. From *Mechanics' Pocket Memoranda: A Convenient Pocketbook,* 6th ed. (Scranton, PA: Colliery Engineer Company, 1900), Appendix.

Page 231
From "Wood-Engravers—A. Lepère," *Scribner's Magazine,* December 1895, 720.

Page 237
How the Wrights cut their first propellers to varying cross-sections, from three spruce boards. From W. Kaempffert, *The New Art of Flying* (New York: Dodd, Mead, 1911), 109.

Page 241
Locomotive evolution as portrayed in a book for young boys. From F. B. Masters, "What Every American Boy—and Man—Should Know About Locomotives," *St. Nicholas,* Vol. XLII (1915), Part 1, 538.

Index

Note: Page numbers in **bold** refer to a chapter or major treatment. Page numbers in *italics* refer to illustrations. Page numbers with a *t* refer to tables.

Index

Index

gestation, 158
of technologies, 160*t*
GI Bill, 230, 231
Gibbs, J. W., 41
Giffard, Henri, 24, 27, 125
Ginzburg, Carlo, 168, 170
Gladwell, Malcolm, 234
glasses, for reading, 150, *151*
gliders, 24, *29*, 29–30
glue, 218
Goldberg, Rube, 71
Goldwater, Barry, 39
Gordon, Flash, vii
Gowallapus, Shadrach, 10
Great Western (steamboat), 106
Greeks, 13, 42
Gregory, Hanson, 9
Guericke, Otto von, 48–49, *49*
gunpowder, 53
Gutenberg, Johann, **138**, 153, 157, 163, 168,
180, 191, 193, 198, 233, 241
Bible of, 140–42, *141, 142*, 163
businesses of, 139
indulgences and, 140
money and, 143
perfectionism of, 142
press of, *163*
type, *167*
Haines, Flora, 30, 31, 32
Hall, Michael, 25
Haller, Albrecht, 79, 80–81, 87
poetry of, *80*
Hamilton, Alexander, 52
Handbook of Nature Science (Haines), 30
Hate, 39, 40
Hayes, Rutherford B., 222
heat, 40
of blood, 82
entropy and, 90
extent of, 76
food and, 87
intensity of, 76
latent, 77, 78
mechanical theory of, 83, 90
observation of, *77*
thermodynamics and, 72
Heat: A Mode of Motion (Tyndall), 91
Heilmann, Andreas, 139, 140
Heisenberg's uncertainty calculation/relation,
121, 233
Hellenism, 42
Henson, William Samuel, 26
herbals, as illustrations, *176, 177*
Hermes Avitor (dirigible), 27
Hermes Jr. Avitor (dirigible), 27
Hero of Alexandria, 44
Hero's turbine, 44, *44*, 46, 55, 95, 112
Hewitt, Abram S., 219–20, *220*, 222
hieroglyphs, 11–12
Hlavacek, Petr, 6
Homo sapiens, 242

Homo technologicus, 242
Hone, Philip, 99
Hooke, Robert, 52, 75
houille blanche, 112
Howley, Mary Anne, 202, *202*, 205, 225
Hubble photograph, *170*
Huguenots, 52, 57
Hulls, Jonathan, 101
steam tug of, *101*
Humphreys, Colin, 160
Huygens, Christian, 51, 52, 60
hydrostatic forces, 60

Ibn Firnas, Abbas, 21, 22, 24
ice, 20
ICS. *See* International Correspondence
Schools
Iliad, 13
illustrations, **173**, *176*
Imhotep, 12, *12*
incunabula, 157, *161*, 165
Industrial Revolution, 57, 98, 198, 211
ingenuity-limited, 121
inks, 164
Inness, George, 99
Inquisition, 46, 169, 171
intaglio, 179, *180*
integrated circuits, 160
International Correspondence Schools (ICS),
227–28, *229*
International System of units, 89
Internet, 160
invention(s)
aggregation of, viii, 238
arc of, **233**, 238, 239
causality of, viii
computers as, 159
counterclaims in, 19
cradle of, 158
dates when motivated begins, 130*t*
as evolution, 138
exponential envelope of, *240*
freedom and, 118
gestation of, 158
individual credit for, 15
Latin proverb, 242
motivation for, **117**, 125
vs. multigenium, 238
myths of, 7–8
necessity and, 118
pleasure and, 118
of priority, **8**
profit and, 117–18
of speed, 100
urgency and, 127
inventor(s)
assignment of, 11
canonical, 19
isentropic expansion, 37

James, William, 210
Jefferson, Thomas, 68,

Index

Index

Index